AF595464

Virus Infection in Equine

Virus Infection in Equine

Editors

Amir Steinman
Oran Erster

MDPI • Basel • Beijing • Wuhan • Barcelona • Belgrade • Manchester • Tokyo • Cluj • Tianjin

Editors
Amir Steinman
Koret School of Veterinary Medicine
The Hebrew University of Jerusalem
Rehovot
Israel

Oran Erster
Central Virology Laboratory
Haim Sheba Medical Center
Ministry of Health
Ramat Gan
Israel

Editorial Office
MDPI
St. Alban-Anlage 66
4052 Basel, Switzerland

This is a reprint of articles from the Special Issue published online in the open access journal *Animals* (ISSN 2076-2615) (available at: www.mdpi.com/journal/animals/special_issues/viie).

For citation purposes, cite each article independently as indicated on the article page online and as indicated below:

LastName, A.A.; LastName, B.B.; LastName, C.C. Article Title. *Journal Name* **Year**, *Volume Number*, Page Range.

ISBN 978-3-0365-5086-2 (Hbk)
ISBN 978-3-0365-5085-5 (PDF)

Cover image courtesy of Sharon Tirosh-Levy

Contents

About the Editors

Amir Steinman

Prof. Amir Steinman, Koret School of Veterinary Medicine, the Hebrew University, is the director of the Koret School of Veterinary Medicine—Veterinary Teaching Hospital. His main research interests are equine infectious diseases and antimicrobial resistance.

Oran Erster

Dr. Oran Erster, Israel Ministry of Health, is the head of the Molecular Diagnostic Unit at the Israel Central Virology Laboratory. His main interests are viral molecular diagnostics, molecular epidemiology of viral diseases, and viral host–pathogen interactions.

MDPI

Editorial

Virus Infection in Equine

Amir Steinman [1,*], Oran Erster [2] and Sharon Tirosh-Levy [1]

1 Koret School of Veterinary Medicine, The Robert H. Smith Faculty of Agriculture, Food and Environment, The Hebrew University of Jerusalem, Rehovot 7610001, Israel; sharontirosh@gmail.com
2 Central Virology Laboratory, Ministry of Health, Haim Sheba Medical Center, Ramat-Gan 5265601, Israel; oran.erster@gmail.com
* Correspondence: amirst@savion.huji.ac.il

Citation: Steinman, A.; Erster, O.; Tirosh-Levy, S. Virus Infection in Equine. *Animals* **2022**, *12*, 957. https://doi.org/10.3390/ani12080957

Received: 2 April 2022
Accepted: 6 April 2022
Published: 8 April 2022

Publisher's Note: MDPI stays neutral with regard to jurisdictional claims in published maps and institutional affiliations.

The relationship between men and horses has significantly evolved over the last century. Throughout history, horses have been used mainly for transportation, work, and war. In developing countries horses still serve as important burden animals for work and transport, while in more developed countries horses are being mainly used for sport, therapeutic riding or kept as pets. Additionally, in some countries, horses also serve as food animals for meat consumption. As a result, there is much interaction between horses and humans, with implications on human health as well as environmental consequences, under the One Health (OH) concept. Horses may serve as reservoirs to various pathogens, transmit certain zoonotic diseases and introduce pathogens into new geographical niches. The recent increase in human and animal international transportation, in combination with global warming and changing environment, has facilitated the spread of numerous infectious diseases into new geographic regions, as clearly demonstrated with the current SARS-CoV-2 outbreak.

One of the most well-known example for the introduction or re-introduction of pathogens to new niches is of West Nile virus (WNV). This mosquito-borne virus is maintained by wild birds and clinically effects mostly humans and horses [1]. Until the mid-1990's, it was not considered an important differential diagnosis for equine and for human neurological disease and was mainly limited to Africa and the Middle East. During the 1990's several large-scale outbreaks in humans and in horses were reported [2–5], including its dramatic appearance in the New York City area in 1999, followed by quick spread throughout the United States of America in the following years [6]. These events drew attention to the potential risk of introduction of pathogens into new global niches, further demonstrating the importance of better surveillance of equine viral diseases. WNV lineage 2 has been spreading with the detection of this lineage in European countries, first in Hungary in 2004 [7], and soon afterwards in neighboring countries [8]. In recent years, WNV lineage 2 has been detected in European countries including Spain in 2018 [9] and more recently in Germany [10]. In 2018, the largest outbreak of human WNV infections in Europe occurred, when 11 countries reported 1548 cases [11]. In the case of WNV, although it cannot be directly transmitted between horses and humans, horses can be used as sentinels for the activity of the virus in the area since they are mostly kept outside and are exposed to mosquitoes bites [1].

Another example for an emerging equine viral disease is Equine Encephalosis virus (EEV), which is reviewed in this special issue [12]. Until 2008, it was considered to be restricted to southern Africa. However, during 2008–2009, EEV was isolated in an outbreak in Israel, demonstrating the introduction of this pathogen into a new niche. In retrospective, this occurred even earlier since antibodies against EEV had already been detected in sera samples that were collected from horses in Israel in 2001 [13]. The spread of EEV from South Africa to central Africa [14], the Middle East [15], and India [16] is another example of the possible emergence of new pathogens in new niches and should be a reminder not to limit the differential diagnoses list when facing a possible outbreak, or a cluster of undiagnosed clinical cases.

Several other viruses have extended their geographical distribution in recent years, among which is Usutu virus (USUV). The apparent incidence of human USUV is very low compared with the incidence of WNV. However, the current knowledge might be biased by the low capacity to correctly identify USUV infection in humans [17]. In a study from Italy, in which cerebrospinal fluid and serum samples were retrospectively analyzed for the presence of USUV antibodies and RNA in human patients, it was concluded that USUV is not a sporadic event in the studied area [18]. In Israel, USUV was only isolated in six pools of mosquitoes in the north of country, five in 2015 and one in 2014 [19], however, until now, no human or equine clinical cases of USUV have been reported. In a survey that was conducted on equine sera samples in 2018, in Israel, 10.8% were seropositive [20], further supporting the possible role of horses as sentinels for the local circulation of such viruses, which might be otherwise underdiagnosed.

Equine viral diseases can be divided into three groups: zoonotic pathogens, viruses with significant veterinary and economic significance, and viruses with minimal clinical impact on horses.

Zoonotic viruses can be further divided to those that can be transmitted directly from horses to humans such as Rabies, Hendra and Venezuelan equine encephalitis (VEE) epizootic strains, and those that affect horses and humans but cannot be directly transmitted between them. These include WNV, USUV, Eastern Equine Encephalitis (EEE), Western Equine Encephalitis (WEE) and others.

Rabies, is an example for a virus that can be directly transmitted from horses to humans, which is unfortunately still endemic in many countries, including Israel. Although horses can be infected, they are not an important source of human rabies infections. Yet, because of its severity, it is highly recommended to vaccinate horses in endemic areas, in order to prevent human infections. In Israel, between 1995 and 2019, rabies was detected in nine horses and in four donkeys [21]. No cases of rabies were detected in neither horses nor donkeys in 2020 [22]. During the same period (1995–2020), 1109 rabies cases were detected in other animals in Israel [22], further demonstrating the limited role of horses in the epidemiology of rabies. Although human cases are extremely rare in Israel, rabies vaccination is mandatory for competing horses and for imported horses and is highly recommended for all other horses.

Viruses that can cause severe economic losses include, for example, Equine Influenza virus (EIV) and Equine alpha Herpesvirus 1 (EHV-1). The economic impact of the EIV outbreak that occurred in Australia in 2007 was enormous, with an estimated financial cost to the governments of hundreds of millions of dollars, and the cost for the horse industry and associated businesses was even higher [23]. From an economic perspective, Equine Herpesviruses (EHV-1 in particular) are one of the most important equine pathogens in Europe [24]. EHV-1 may cause significant economic losses due to its high morbidity and the restrictions that are required to limit its spread. In three studies published in this special issue, the authors have evaluated the exposure of different cohorts to this important pathogen [25–27]. African Horse Sickness should also be included in this group and although it is still limited mainly to Africa, measures should be implemented to prevent its spread, which should include active surveillance as well as preventive vaccination in neighboring countries.

The third group of viral pathogens include those that may cause mild to moderate disease to limited number of horses, and viruses that may not cause clinical signs and remain asymptomatic in horses. Several such viruses were recently found in horses, by either serology or by PCR, but their clinical significance is yet unknown. These include Equine Corona virus (ECoV) [28], Equine Parvoviruses [29], Equine Herpes virus-2, 5 [27] and Equine Encephalosis virus [12], which are described in this special issue.

The overall aim of this Special Issue is to provide updated information on different aspects of equine viral diseases, including their prevalence, pathogenesis and diagnostics in different cohorts.

El Brini et al. evaluated the seroprevalence of EHV-1 and EHV-4 among horses in the north of Morocco, as well as the antibody titers in vaccinated horses under field conditions. Overall, 12.8% unvaccinated and 21.8% vaccinated horses were positive to EHV-1 and all were positive to EHV-4 demonstrating that both viruses are endemic in the north of Morocco, with prevalence differences between regions. Furthermore, horses vaccinated with a monovalent EHV-1 vaccine had low antibody titers, which needs to be addressed [26].

El-Hage et al. aimed to better understand the role of EHVs infection in horses in Victoria, Australia, with and without clinical signs of respiratory disease, by PCR. Whereas only few horses were PCR positive for EHV-1 (three horses) and EHV-4 (five horses), many were PCR positive for EHV-2 (20.3%) and EHV-5 (60.2%). Although the odds of EHV-5 positive horses, demonstrating clinical signs of respiratory disease, were twice that of EHV-5 negative horses, no quantitative difference between mean loads of EHV shedding was found between the two groups and the clinical significance of respiratory gammaherpesvirus infections in horses is still unclear [27].

Bazanow et al. estimated the serological status of a semi-isolated group of horses (Huculs) in Poland, by using nasal secretions and sera samples. All the nasal swabs were negative for the tested viruses. Among the 20 horses that were tested, antibodies were detected against EHV-1 in 12 horses (60%), EIV A/H7N7 in 13 (65%), EIV A/H3N8 in 12 (60%), USUV in five (25%), and Equine Rhinitis A virus (ERAV) in one (5%), whereas antibodies against Equine Arteritis virus (EAV), Equine Infectious Anaemia virus (EIAV), and WNV were not detected. These results indicates that the Hucul herd could be used as sentinels for the detection of equine pathogens in the selected area [25].

Limited information is also available regarding Equine Parvoviruses, which were only recently identified. Pusterla et al. have tested the molecular prevalence of three Parvoviruses in blood and respiratory secretions of sick and healthy horses in the USA. Equine Parvoviruses were detected in both sick and healthy horses in similar frequencies suggesting that their role in equine respiratory disease is limited and should be further explored [29].

Schvartz et al. evaluated the risk of exposure to Equine Corona virus (ECoV) of horses in Israel. Exposure to ECoV was detected in 17 of 29 farms (58.6%) and in 41 horses (12.3%). The geographical area was the only factor that was found to be significantly associated with ECoV exposure. The results of this study indicate that ECoV should be included in the differential diagnosis list of pathogens in cases of adult horses with relevant clinical signs [28].

Lawton et al. have tested nasal secretions from equids with acute onset of fever and respiratory signs using qPCR and sera samples from healthy horses with possible exposure to humans with SARS-CoV-2 infection in the USA using ELISA. None of the clinical horses was found positive for SARS-CoV-2, whereas 35/587 (5.9%) apparently healthy Thoroughbred racing horses had detectable IgG antibodies to SARS-CoV-2. The authors have concluded that while horses appear to be susceptible to SARS-CoV-2 when in close contact with infected humans, they do not seem to be clinically affected [30].

Two reviews are included in this special issue, the first by Knox and Beddoe who reviewed current isothermal diagnostic techniques available for the detection of equine viruses of zoonotic concern, and provide insight into their potential for in-field deployment [31]. The second by Tirosh-Levy and Steinman who summarizes current knowledge regarding EEV structure, pathogenesis, clinical significance, and epidemiology [12].

Increased international transportation and trade over the last few decades have increased the risk of the introduction of pathogens into new areas. Global climate change has influenced environmental conditions and the ability of pathogens to survive, as well as changed the habitats of certain vectors and hosts. These processes have led to the emergence or re-emergence of various pathogens in different parts of the world, including those affecting horses. This special issue has featured some aspects regarding several well recognized as well as some new and emerging equine viral pathogens, highlighting the need of updated epidemiological data. Such surveillance is crucial for proper decision making by clinicians and by regulatory authorities. As well demonstrated by the recent global emergence of SARS-CoV-2, the development of an effective infrastructure for the rapid and effective detection and control of novel viral pathogens, as well as re-emerging ones is essential. Horses should play an important role in such surveillance system, not only for equine pathogens but also as sentinels to other viruses and arboviruses. As was demonstrated in several examples in this special issue, it is important to remember both as clinicians and as researchers, that when facing clinical cases, even when those are common, we should remain alert to the possibility of the intrusion of unknown pathogens and, therefore, should seek for a definitive diagnosis. This may allow for early detection of emerging or re-emerging pathogens.

Author Contributions: Conceptualization, A.S.; writing—original draft preparation, A.S. and S.T.-L.; writing—review and editing, A.S., O.E. and S.T.-L. All authors have read and agreed to the published version of the manuscript.

Funding: This research received no external funding.

Institutional Review Board Statement: Not applicable.

Informed Consent Statement: Not applicable.

Data Availability Statement: Not applicable.

Conflicts of Interest: The authors declare no conflict of interest.

References

1. Venter, M.; Pretorius, M.; Fuller, J.A.; Botha, E.; Rakgotho, M.; Stivaktas, V.; Weyer, C.; Romito, M.; Williams, J. West Nile Virus Lineage 2 in Horses and Other Animals with Neurologic Disease, South Africa, 2008–2015. *Emerg. Infect Dis.* **2017**, *23*, 2060–2064. [CrossRef] [PubMed]
2. Murgue, B.; Murri, S.; Triki, H.; Deubel, V.; Zeller, H.G. West Nile in the Mediterranean basin: 1950–2000. *Ann. N. Y. Acad. Sci.* **2001**, *951*, 117–126. [CrossRef] [PubMed]
3. Cantile, C.; Di Guardo, G.; Eleni, C.; Arispici, M. Clinical and neuropathological features of West Nile virus equine encephalomyelitis in Italy. *Equine Vet. J.* **2000**, *32*, 31–35. [CrossRef] [PubMed]
4. Murgue, B.; Murri, S.; Zientara, S.; Durand, B.; Durand, J.P.; Zeller, H. West Nile outbreak in horses in southern France, 2000: The return after 35 years. *Emerg. Infect. Dis.* **2001**, *7*, 692–696. [CrossRef]
5. Steinman, A.; Banet, C.; Sutton, G.A.; Yadin, H.; Hadar, S.; Brill, A. Clinical signs of West Nile virus encephalomyelitis in horses during the outbreak in Israel in 2000. *Vet. Rec.* **2002**, *151*, 47–49. [CrossRef]
6. Gerhardt, R. West Nile virus in the United States (1999–2005). *J. Am. Anim. Hosp. Assoc.* **2006**, *42*, 170–177. [CrossRef]
7. Bakonyi, T.; Ivanics, E.; Erdelyi, K.; Ursu, K.; Ferenczi, E.; Weissenbock, H.; Nowotny, N. Lineage 1 and 2 strains of encephalitic West Nile virus, central Europe. *Emerg. Infect. Dis.* **2006**, *12*, 618–623. [CrossRef]
8. Bergmann, F.; Trachsel, D.S.; Stoeckle, S.D.; Bernis Sierra, J.; Lübke, S.; Groschup, M.H.; Gehlen, H.; Ziegler, U. Seroepidemiological Survey of West Nile Virus Infections in Horses from Berlin/Brandenburg and North Rhine-Westphalia, Germany. *Viruses* **2022**, *14*, 243. [CrossRef]
9. Napp, S.; Llorente, F.; Beck, C.; Jose-Cunilleras, E.; Soler, M.; Pailler-García, L.; Amaral, R.; Aguilera-Sepúlveda, P.; Pifarré, M.; Molina-López, R. Widespread Circulation of Flaviviruses in Horses and Birds in Northeastern Spain (Catalonia) between 2010 and 2019. *Viruses* **2021**, *13*, 2404. [CrossRef]
10. Ziegler, U.; Luhken, R.; Keller, M.; Cadar, D.; van der Grinten, E.; Michel, F.; Albrecht, K.; Eiden, M.; Rinder, M.; Lachmann, L.; et al. West Nile virus epizootic in Germany, 2018. *Antivir. Res.* **2019**, *162*, 39–43. [CrossRef]
11. Bakonyi, T.; Haussig, J.M. West Nile virus keeps on moving up in Europe. *Euro Surveill* **2020**, *25*, 2001938. [CrossRef] [PubMed]
12. Tirosh-Levy, S.; Steinman, A. Equine Encephalosis Virus. *Animals* **2022**, *12*, 337. [CrossRef] [PubMed]
13. Westcott, D.; Wescott, D.G.; Mildenberg, Z.; Bellaiche, M.; McGowan, S.L.; Grierson, S.S.; Choudhury, B.; Steinbach, F. Evidence for the circulation of equine encephalosis virus in Israel since 2001. *PLoS ONE* **2013**, *8*, e70532. [CrossRef] [PubMed]
14. Oura, C.A.; Batten, C.A.; Ivens, P.A.; Balcha, M.; Alhassan, A.; Gizaw, D.; Elharrak, M.; Jallow, D.B.; Sahle, M.; Maan, N.; et al. Equine encephalosis virus: Evidence for circulation beyond southern Africa. *Epidemiol. Infect.* **2012**, *140*, 1982–1986. [CrossRef] [PubMed]

15. Tirosh-Levy, S.; Gelman, B.; Zivotofsky, D.; Quraan, L.; Khinich, E.; Nasereddin, A.; Abdeen, Z.; Steinman, A. Seroprevalence and risk factor analysis for exposure to equine encephalosis virus in Israel, Palestine and Jordan. *Vet. Med. Sci.* **2017**, *3*, 82–90. [CrossRef]
16. Yadav, P.D.; Albarino, C.G.; Nyayanit, D.A.; Guerrero, L.; Jenks, M.H.; Sarkale, P.; Nichol, S.T.; Mourya, D.T. Equine Encephalosis Virus in India, 2008. *Emerg. Infect. Dis.* **2018**, *24*, 898–901. [CrossRef]
17. Luhken, R.; Jost, H.; Cadar, D.; Thomas, S.M.; Bosch, S.; Tannich, E.; Becker, N.; Ziegler, U.; Lachmann, L.; Schmidt-Chanasit, J. Distribution of Usutu Virus in Germany and Its Effect on Breeding Bird Populations. *Emerg. Infect. Dis.* **2017**, *23*, 1994–2001. [CrossRef]
18. Grottola, A.; Marcacci, M.; Tagliazucchi, S.; Gennari, W.; Di Gennaro, A.; Orsini, M.; Monaco, F.; Marchegiano, P.; Marini, V.; Meacci, M.; et al. Usutu virus infections in humans: A retrospective analysis in the municipality of Modena, Italy. *Clin. Microbiol. Infect.* **2017**, *23*, 33–37. [CrossRef]
19. Mannasse, B.; Mendelson, E.; Orshan, L.; Mor, O.; Shalom, U.; Yeger, T.; Lustig, Y. Usutu Virus RNA in Mosquitoes, Israel, 2014–2015. *Emerg. Infect. Dis.* **2017**, *23*, 1699–1702. [CrossRef]
20. Schvartz, G.; Tirosh-Levy, S.; Erester, O.; Shenhar, R.; Levy, H.; Bazanow, B.; Gelman, B.; Steinman, A. Exposure of Horses in Israel to West Nile Virus and Usutu Virus. *Viruses* **2020**, *12*, 1099. [CrossRef]
21. Balaish, M. Veterinary Annual Reports. Israeli Veterinary Services and Animal Health 2019. Available online: https://www.gov.il/he/departments/publications/reports/doch_shnati_2019 (accessed on 6 March 2022).
22. Balaish, M. Veterinary Annual Reports. Israeli Veterinary Services and Animal Health 2020. Available online: https://www.gov.il/he/departments/publications/reports/doch_shnati_2020 (accessed on 6 March 2022).
23. Smyth, G.; Dagley, K.; Tainsh, J. Insights into the economic consequences of the 2007 equine influenza outbreak in Australia. *Aust. Vet. J.* **2011**, *89*, 151–158. [CrossRef] [PubMed]
24. Lecollinet, S.; Pronost, S.; Coulpier, M.; Beck, C.; Gonzalez, G.; Leblond, A.; Tritz, P. Viral equine encephalitis, a growing threat to the horse population in Europe? *Viruses* **2019**, *12*, 23. [CrossRef] [PubMed]
25. Bażanów, B.; Pawęska, J.T.; Pogorzelska, A.; Florek, M.; Frącka, A.; Gębarowski, T.; Chwirot, W.; Stygar, D. Serological Evidence of Common Equine Viral Infections in a Semi-Isolated, Unvaccinated Population of Hucul Horses. *Animals* **2021**, *11*, 2261. [CrossRef] [PubMed]
26. El Brini, Z.; Fassi Fihri, O.; Paillot, R.; Lotfi, C.; Amraoui, F.; El Ouadi, H.; Dehhaoui, M.; Colitti, B.; Alyakine, H.; Piro, M. Seroprevalence of Equine Herpesvirus 1 (EHV-1) and Equine Herpesvirus 4 (EHV-4) in the Northern Moroccan Horse Populations. *Animals* **2021**, *11*, 2851. [CrossRef]
27. El-Hage, C.; Mekuria, Z.; Dynon, K.; Hartley, C.; McBride, K.; Gilkerson, J. Association of Equine Herpesvirus 5 with Mild Respiratory Disease in a Survey of EHV1,-2,-4 and-5 in 407 Australian Horses. *Animals* **2021**, *11*, 3418. [CrossRef]
28. Schvartz, G.; Tirosh-Levy, S.; Barnum, S.; David, D.; Sol, A.; Pusterla, N.; Steinman, A. Seroprevalence and risk factors for exposure to equine coronavirus in apparently healthy horses in israel. *Animals* **2021**, *11*, 894. [CrossRef]
29. Pusterla, N.; James, K.; Barnum, S.; Delwart, E. Investigation of Three Newly Identified Equine Parvoviruses in Blood and Nasal Fluid Samples of Clinically Healthy Horses and Horses with Acute Onset of Respiratory Disease. *Animals* **2021**, *11*, 3006. [CrossRef]
30. Lawton, K.O.; Arthur, R.M.; Moeller, B.C.; Barnum, S.; Pusterla, N. Investigation of the Role of Healthy and Sick Equids in the COVID-19 Pandemic through Serological and Molecular Testing. *Animals* **2022**, *12*, 614. [CrossRef]
31. Knox, A.; Beddoe, T. Isothermal Nucleic Acid Amplification Technologies for the Detection of Equine Viral Pathogens. *Animals* **2021**, *11*, 2150. [CrossRef]

Article

Investigation of the Role of Healthy and Sick Equids in the COVID-19 Pandemic through Serological and Molecular Testing

Kaila O. Y. Lawton [1], Rick M. Arthur [2], Benjamin C. Moeller [3,4], Samantha Barnum [1] and Nicola Pusterla [1,*]

1 Department of Medicine and Epidemiology, School of Veterinary Medicine, University of California, Davis, CA 95616, USA; kolawton@ucdavis.edu (K.O.Y.L.); smmapes@ucdavis.edu (S.B.)
2 School of Veterinary Medicine, University of California, Davis, CA 95616, USA; rmarthur@ucdavis.edu
3 KL Maddy Equine Analytical Chemistry Laboratory, School of Veterinary Medicine, University of California, Davis, CA 95616, USA; bcmoeller@ucdavis.edu
4 Department of Molecular Biosciences, School of Veterinary Medicine, University of California, Davis, CA 95616, USA
* Correspondence: npusterla@ucdavis.edu; Tel.: +1-530-752-1039

Simple Summary: The objective of the present study was to determine if horses are susceptible to SARS-CoV-2. Nasal swabs from 667 equids with acute onset of fever and respiratory signs were tested by qPCR for SARS-CoV-2. Further, 633 serum samples collected from a cohort of 587 healthy racing Thoroughbreds with possible exposure to humans with SARS-CoV-2 infection were tested for antibodies to SARS-CoV-2 using an ELISA targeting the receptor-binding domain of the spike protein. All 667 horses with fever and respiratory signs tested qPCR-negative for SARS-CoV-2. A total of 35/587 (5.9%) Thoroughbred racing horses had detectable IgG antibodies to SARS-CoV-2. While horses appear to be susceptible to SARS-CoV-2 when in close contact with humans with SARS-CoV-2 infection, clinical disease was not observed in the study horses. Experimental challenge studies using pure inocula are needed in order to study the clinical, hematological, molecular, and serological features of adult horses infected with SARS-CoV-2.

Abstract: More and more studies are reporting on the natural transmission of SARS-CoV-2 between humans with COVID-19 and their companion animals (dogs and cats). While horses are apparently susceptible to SARS-CoV-2 infection based on the homology between the human and the equine ACE-2 receptor, no clinical or subclinical infection has yet been reported in the equine species. To investigate the possible clinical role of SARS-CoV-2 in equids, nasal secretions from 667 horses with acute onset of fever and respiratory signs were tested for the presence of SARS-CoV-2 by qPCR. The samples were collected from January to December of 2020 and submitted to a commercial molecular diagnostic laboratory for the detection of common respiratory pathogens (equine influenza virus, equine herpesvirus-1/-4, equine rhinitis A and B virus, *Streptococcus equi* subspecies *equi*). An additional 633 serum samples were tested for antibodies to SARS-CoV-2 using an ELISA targeting the receptor-binding domain of the spike protein. The serum samples were collected from a cohort of 587 healthy racing Thoroughbreds in California after track personnel tested qPCR-positive for SARS-CoV-2. While 241/667 (36%) equids with fever and respiratory signs tested qPCR-positive for at least one of the common respiratory pathogens, not a single horse tested qPCR-positive for SARS-CoV-2. Amongst the racing Thoroughbreds, 35/587 (5.9%) horses had detectable antibodies to SARS-CoV-2. Similar to dogs and cats, horses do not seem to develop clinical SARS-CoV-2 infection. However, horses can act as incidental hosts and experience silent infection following spillover from humans with COVID-19. SARS-CoV-2-infected humans should avoid close contact with equids during the time of their illness.

Keywords: SARS-CoV-2; horses; nasal secretions; blood; qPCR; ELISA; sick equids; healthy horses

Citation: Lawton, K.O.Y.; Arthur, R.M.; Moeller, B.C.; Barnum, S.; Pusterla, N. Investigation of the Role of Healthy and Sick Equids in the COVID-19 Pandemic through Serological and Molecular Testing. *Animals* **2022**, *12*, 614. https://doi.org/10.3390/ani12050614

Academic Editors: Margarita Martín Castillo and Harold C. McKenzie III

Received: 21 December 2021
Accepted: 25 February 2022
Published: 28 February 2022

1. Introduction

Epidemiological work in the field of SARS-CoV-2 has focused on the human–animal interface in order to identify animal species, which could act as reservoirs and intermediate hosts [1]. Understanding the host range for SARS-CoV-2 is important in order to control the ongoing pandemic and to protect populations of wild and domestic animals in their native habitat and under human care, respectively. The best-documented evidence for susceptibility of any animal species comes from detecting SARS-CoV-2 under natural conditions or proof of active viral transmission between infected and susceptible in contact animals. While experimental inoculations of selected animal species are needed to document viral kinetics and risk of viral transmission, such protocols only mirror, but never reproduce, natural conditions. The predictive susceptibility of animals has also been based on computational modelling of their angiotensin-I-converting enzyme 2 (ACE-2), a key receptor for SARS-CoV-2 [2]. ACE-2 serves as a functional receptor for the spike protein of SARS-CoV-2 [3]. Cross-species infections can occur when a coronavirus adapts to a new host in part through the mutation of the spike protein, shown to enhance the binding affinity for ACE-2 [4]. Using comparative genomic approaches and protein structural analysis, Damas and colleagues [2] determined the conservation of ACE-2 and its potential to be used as a receptor by SARS-CoV-2 in 410 vertebrate species. Their results showed that mammals fell into low to high binding categories, with *equus caballus* and *equus asinus* displaying a low binding score category for SARS-CoV-2.

The close interactions of domestic animals with humans worldwide make determining their susceptibility an urgent need. Human-to-animal transmissions of SARS-CoV-2 have been documented in dogs, cats, tigers, lions and minks [5–7]. The role of equids in the COVID-19 pandemic has remained poorly investigated. Horses are potentially susceptible to SARS-CoV-2 based on the binding affinity and stability between ACE-2 and the receptor-binding domain of the S protein [8,9]. Considering the large number of equids globally and the direct or indirect contact these animals have with humans, information pertaining to their susceptibility to SARS-CoV-2 and their role in virus transmission is needed. Therefore, the aims of the present study were to determine if SARS-CoV-2 could be detected in nasal secretions of equids with acute onset of fever and respiratory signs using qPCR and to investigate the seroprevalence against SARS-CoV-2 in a cohort of racing horses with possible exposure to humans with SARS-CoV-2 infection.

2. Materials and Methods

2.1. Study Population and Sampling

Nasal fluid samples from 667 equids with acute onset of upper airway infection were enrolled in the study. The same samples were used to investigate three newly identified equine parvoviruses in a recent study [10]. The respiratory secretions were submitted to a commercial diagnostic laboratory from 1 January 2020 to 31 December 2020 for the molecular detection of common respiratory pathogens, including equine influenza virus (EIV), equine herpesvirus-1/-4 (EHV-1/-4), equine rhinitis A and B virus (ERVs) and *Streptococcus equi* subspecies *equi* (*S. equi*).

Six hundred and thirty-three serum samples from 587 racing Thoroughbred horses from California, collected from 10 July 2020 to 12 September 2020, were available for antibody testing against SARS-CoV-2. The blood samples had been collected as part of the routine medication testing program established by the California Horse Racing Board. For the majority of the racing horses, only one serum sample was available, while 2 and 3 consecutive serum samples were available for 36 and 5 horses, respectively. The samples were stored at −80 °C until testing. The period of sample collection coincided with a known outbreak of COVID-19 at the sampling location with 22 asymptomatic track personnel testing qPCR-positive for SARS-CoV-2 (https://www.washingtonpost.com/sports/2020/07/16/del-mar-cancels-racing-after-22-positive-covid-19-tests-among-jockeys-track-workers, accessed on 1 November 2021). Because of confidentiality issues, only the age and sex of the

587 racing Thoroughbred horses were made available to the researchers performing the testing by sample identification numbers.

Serum samples collected from 88 healthy adult horses in 2015 (pre-COVID-19 pandemic) and stored at −80 °C until testing served as negative control to establish the cutoff value for the ELISA. Serum samples from 24 horses with previously confirmed ECoV infection were available to test possible cross-reactivity using the SARS-CoV-2 ELISA [11].

2.2. Quantitative PCR Analyses

Nasal fluid samples from 667 horses with acute onset of fever and respiratory signs were tested for the presence of EIV, EHV-1/-4, ERVs and *S. equi* as previously reported [10,12,13]. Primers and probes targeting the S gene of SARS-CoV-2 were designed following BLAST analysis of published sequences from GenBank (www.ncbi.nlm.nih.gov/genbank, accessed on 1 March 2020) (Table 1). Amplification of the target gene was performed using a commercial thermocycler/fluorometer (QuantStudio 5, Applied Biosystems, Foster City, CA, USA). The standard amplification conditions were as follows: 2 min at 50 °C, 10 min at 95 °C, and 40 cycles of 15 s at 95 °C and 60 s at 60 °C. Each PCR reaction for the 6 equine respiratory pathogens and SARS-CoV-2 contained a commercially available mastermix (Universal TaqMan Mastermix with AmpErase UNG, Applied Biosystems, Foster City, CA, USA), 0.625 U of AmpliTaq Gold, 400 nM of each primer and 80 nM of the respective TaqMan probe, and 1 µL of DNA or 5 µL of cDNA sample for a total volume of 12 µL. For the SARS-CoV-2 qPCR assay, a standard curve was generated using plasmid containing the target sequence (Table 1). The amplification efficiency of the SARS-CoV-2 qPCR assay was calculated from the slope using the formula E = 10^(−1/slope). The amplification efficiency was 99% for the spike protein gene of SARS-CoV-2, indicating a very high analytical sensitivity. The detection limit for the SARS-CoV-2 qPCR assay was 13 genome equivalents when the cDNA was purified from nasal secretions. The quality and efficiency of nucleic acid extraction were determined by targeting an equine housekeeping gene as previously described [12].

Table 1. Oligonucleotide sequences of primers, probe and positive plasmid control used to detect SARS-CoV-2 by qPCR.

Target Gene (GenBank)	Oligonucleotides
Spike gene (MT773134)	SARS-CoV-2-forward primer: GGCACAGGTGTTCTTACTGAGTCTAAC SARS-CoV-2-reverse primer: CAAGTGTCTGTGGATCACGGAC SARS-CoV-2-probe: FAM-TGGCAGAGACATTGCTGA-MGB Plasmid positive control: TTCAACTTCAATGGTTTAACAGGCACAG GTGTTCTTA CTGAGTCTAACAAAAAGTTTCTGCCTTTCCAACAAT TTGGCAGAGACATTGCTGACAC-TACTGATGCTGTCCGTGATCCACAGACACTTGAGATTCTTGACATTACACCATGT

2.3. Serology

Antibody detection was performed by adapting an assay initially described by Zhao and colleagues [14]. The assay targets the S protein, specifically the immunodominant receptor-binding domain (RBD). Microtiter plates were coated with 100 µL of recombinant SARS-CoV-2 RBD of the spike protein (ThermoFisher Scientific, Waltham, MA, USA) diluted in coating buffer (Bethyl Laboratories Inc., Montgomery, TX, USA) at a concentration of 100 ng/mL. Plates were then covered and stored at 4 °C overnight. Serum samples from the study horses previously stored at −80 °C were thawed overnight at 4 °C. On the day of the analysis, the coated plates were washed 4 times with 200 µL of wash buffer (Bethyl Laboratories Inc., Montgomery, TX, USA) per well and gently tapped until dry. Then, each well received 90 µL of sample dilution buffer (Bethyl Laboratories Inc., Montgomery, TX, USA) and 10 µL of serum; each sample was run in singlet. Optimal S protein and serum dilutions were determined prior to assay validation using standard checkboard titration procedures. After the serum samples were loaded into the wells, the plates were covered and wrapped in aluminum foil and incubated for 2 h at room temperature on a titer plate shaker. Thereafter, the plates were washed 4 times, and 100 μL of diluted anti-horse IgG horseradish peroxidase conjugate (dilution of 1:120,000 in 2% milk; Sigma Aldrich, St. Louis, MO, USA) was added. This step was followed by 1 h incubation as mentioned above. After washing the plate 4 times, 100 μL of enzyme substrate (Bethyl Laboratories Inc., Montgomery, TX, USA) was added to each well. The plate was then incubated at room temperature for 10 min. As a final step, 50 μL of stop solution (4.89 mL of 98% sulfuric acid diluted with 495 mL of distilled water) was added to each well. The optical density (OD) was measured at 450 nm in a microplate photometer (Spectramax 250, Molecular Devices Corp., Sunnyvale, CA, USA). The OD was measured within 15 min of adding the stop solution. Cut-off values were determined as six times the standard deviations above the mean value of reactivity of 88 seronegative samples from a pre-COVID-19 cohort of healthy adult horses [15]. Because of the inability to test the serum samples using the reference standard of virus neutralization, seropositive serum samples determined via the ELISA targeting the SARS-CoV-2 RBD of the spike protein were defined as suspect positive.

2.4. Statistical Analyses

Demographic and clinical information from horses with upper airway infection, healthy racing horses and healthy controls was evaluated using descriptive analyses. All statistical analyses were performed using Stata Statistical Software (College Station, TX, USA), and statistical significance was set at $p < 0.05$.

3. Results

Demographic and clinical information from horses with acute onset of fever and respiratory signs was previously reported [10]. Briefly, the population ranged in age from 1 month to 34 years (median 9 years), with greater numbers of males (61%) compared to females (39%). A variety of breeds were represented and included Quarter Horse (37%), Warmblood (14%), Thoroughbred (10%), pony breed (6%), Arabian (5%), Paint Horse (4%) and other breeds (22%). The three most commonly reported clinical signs included fever (97%, range 38.6 to 41.4 °C, median 39.4 °C), nasal discharge (74%) and coughing (46%). Common respiratory pathogens were detected in 241/667 (36%) sick equids (81 EIV, 61 *S. equi*, 50 EHV-4, 36 ERVs, 13 EHV-1). Overall, not a single equid tested qPCR-positive for SARS-CoV-2 by qPCR.

The 88 pre-COVID-19 control horses were composed of 53 males (60%) and 35 females (40%) ages 2 to 12 years (median 4.6 years). The OD for the 88 pre-COVID-19 horses ranged from 0.030 to 0.358 (median 0.122, Figure 1). The cutoff value for a suspect positive SARS-CoV-2 ELISA was set at an OD value of $\geq$0.507. The population of ECoV-seropositive horses was composed of 13 males (54%) and 11 females (46%) aged 4–22 years (median 17.5 years). All 24 ECoV-seropositive horses were seronegative for SARS-CoV-2 with OD values ranging from 0.081 to 0.384 (median 0.137). The population of 587 racing horses was composed of 335 males (57%) and 252 females (43%) aged 2–7 years (median 3 years). The OD for the 633 serum samples ranged from 0.004 to 1.298 (median 0.091). A total of 40/633 serum samples (6.3%) were considered suspect seropositive for SARS-CoV-2 by ELISA with an OD $\geq$ 0.507 (range 0.510 to 1.298, median 0.911; Figure 1). The 40 SARS-CoV-2 suspect seropositive serum samples originated from 35/587 horses (5.9%). Thirty-one horses had a single SARS-CoV-2 suspect seropositive sample, three horses had two suspect seropositive samples (days between serum collections ranged from 28 to 44 days) and one horse had three suspect seropositive samples (days from first to third serum collection was 46 days). Amongst the thirty-one horses with a single SARS-CoV-2 suspect seropositive sample, four horses showed seroconversion between two sample collection time points (days between serum collections ranged from 22 to 41 days).

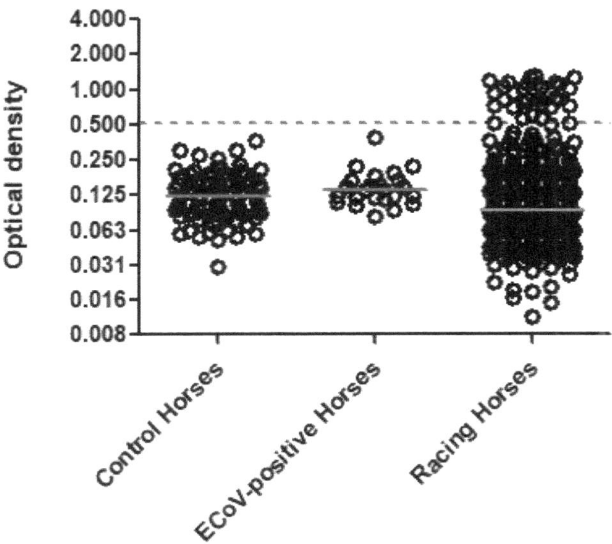

Figure 1. ELISA results from 88 pre-COVID-19 control horses, 24 ECoV-seropositive horses and 633 serum samples collected from 587 racing Thoroughbreds against the recombinant receptor-binding domain of SARS-CoV-2 spike protein. The dashed red line represents the cut-off (0.507). The solid red lines represent the median OD.

4. Discussion

It has been shown that various domestic animal species, including cats, dogs and farmed minks, are susceptible to SARS-CoV-2 infection under natural and experimental conditions [16]. While most of these animal species are permissive to infection, clinical pathology does not always mimic disease observed in humans. Many factors, including genetic diversity, age, comorbidity, expression of ACE-2 receptor and pre-existing diseases, have been shown to modulate disease form [17,18]. Little is known about the prevalence of SARS-CoV-2 in large domestic animal species such as equids. In a serological survey of SARS-CoV-2 in different species of animals from China, no antibodies specific to SARS-CoV-2 were found in serum samples from 18 horses [19]. Therefore, the aim of this study was to investigate the susceptibility to SARS-CoV-2 in equids with acute respiratory disease and in healthy racehorses in close contact with humans with asymptomatic SARS-CoV-2.

The lack of detectable SARS-CoV-2 by qPCR in nasal secretions of 667 horses with acute onset of fever and respiratory signs is in agreement with an investigation performed by IDEXX Reference Laboratories on over 6000 canine, feline and equine specimens tested for SARS-CoV-2 by qPCR from mid-February to mid-April, 2020 (https://www.idexx.com/en/veterinary/reference-laboratories/overview-idexx-sars-cov-2-covid-19-realpcr-test, accessed on 1 November 2021). A recent study evaluating nasal and nasopharyngeal swabs and feces from 34 healthy Italian Trotters with recent contact with SARS-CoV-2 breeders showed no detection of SARS-CoV-2 by qPCR [20]. Another study evaluating the susceptibility of common domestic livestock showed no clinical disease, no nasal and fecal viral shedding determined by qPCR and no virus isolation from respiratory tissues in a single horse following intranasal administration of 6.3 log_{10} plaque-forming units SARS-CoV-2 virus strain 2019-nCoV/USA-WA1/2020 [21]. The reason for negative SARS-CoV-2 qPCR results in the present study population may relate to the lack of disease expression in equids, similar to other domestic animals [6]. Various studies have demonstrated that SARS-CoV-2 infection in companion animals (dogs and cats) is mostly detected in animals living in households with at least one SARS-CoV-2-infected human. The reported frequencies of SARS-CoV-2 infection in dogs and cats confirmed by molecular methods ranges from 0–28% and 0–40%, respectively [22–27]. Close contact of dogs and cats with their SARS-CoV-2-infected owners, especially sharing the bed with an infected human, was recently determined as the main risk factor for transmission [25]. Contact between equids and owners, trainers and barn workers is generally limited in time, with greater physical distances kept between handlers and horses, and contact often occurs in the outdoors. The latter management and husbandry practices are less likely to promote SARS-CoV-2 transmission between SARS-CoV-2-infected humans and equids. To study the impact of SARS-CoV-2-infected horse owners on their horses, prospective longitudinal studies are needed in order to sample horses at regular intervals once horse owners have been diagnosed with COVID-19.

Studies focusing on animals with possible exposure to people with COVID-19 have the potential to quantify the risk of transmission between humans shedding SARS-CoV-2 and susceptible animals. The known asymptomatic qPCR-positive test results of track personnel for SARS-CoV-2 at the racing location represented a unique opportunity to determine potential spillover from infected humans to race horses. The 633 convenience blood samples were collected over a 9-week period, covering a period when racing was cancelled due to the human positive cases. The study results showed that 5.9% of tested horses had antibodies against the RBD of SARS-CoV-2. Due to the small volume of serum available for each racehorse, the samples were run in singlets and the results could not be confirmed via retesting. Further, another limitation was the inability to confirm ELISA positive results using the reference standard of virus neutralization. These limitations may have impacted true seroprevalence against SARS-CoV-2. This relatively high percentage of suspect seropositivity in horses could be related to the large number of infected jockeys and track workers having contact with the racing horses. Of interest was the observation that 4 Thoroughbred racing horses seroconverted to SARS-CoV-2 during the study period. However, the study design does not allow for the determination of whether human-to-horse or horse-to-horse transmission occurred. Nevertheless, to the authors' knowledge, this is the first report showing the exposure of horses with SARS-CoV-2 secondary to spillover from asymptomatic humans. Laboratory-based qPCR is the recommended test for diagnoses of acute cases, while serological tests are important to define epidemiological questions, such as exposure rate [28]. The serological platform used for this study was based on the detection of the RBD of SARS-CoV-2, shown to be one of the most specific antigens [14,29]. Further, the RBD-specific SARS-CoV-2 did not show any cross-reactivity with the closely-related ECoV, ruling out any false-positive results. Studies assessing seroprevalence in companion animals living in households with SARS-CoV-2-infected owners reported seropositivity rates of 3.4–23.5% for dogs and 4–43.8% for cats [23–25,27,30,31]. Because the SARS-CoV-2 shedding status of jockeys and track workers attending every single study horse was unknown, it was impossible to determine the time of infection. Experimental studies using susceptible animals such as cats and documented cat-to-cat transmissions have shown seroconversion occurring as early as 11–12 days post-infection [32]. A similar time to seroconversion can be assumed for other susceptible animal species such as equids. Limitations of the study relate to the lack of longitudinal data from the same horses during the study period, as well as the inability to test nasal or nasopharyngeal secretions for SARS-CoV-2 by qPCR. Further, without sequence information of the SARS-CoV-2 involved in horse and human infections, the authors cannot conclude that horses were infected with the same virus responsible for asymptomatic COVID-19 in humans.

During the monitoring period, no outbreak of a respiratory disease was reported in the racing horses, suggesting that horses with antibodies to SARS-CoV-2 likely experienced subclinical infection. Horses do apparently remain subclinical following infection with SARS-CoV-2. The susceptibility to developing COVID-19 in companion animals is a complex interplay between various viral and host factors [33]. While data is limited on the susceptibility to SARS-CoV-2 infection in domestic animals, it appears that equids are incidental hosts because of occasional SARS-CoV-2 spillover from humans. However, continuous surveillance is necessary in order to monitor the possible transmission of SARS-CoV-2 infection in equids. From a biosecurity perspective, it is highly recommended that humans with clinical and asymptomatic SARS-CoV-2 infection avoid close contact with any companion animals.

5. Conclusions

In conclusion, our results show that equids are susceptible to natural SARS-CoV-2 infections. While SARS-CoV-2 could not be detected via qPCR in nasal secretions of horses with acute onset of fever and respiratory signs, antibodies specific to SARS-CoV-2 were found in 5.9% of healthy racing Thoroughbreds in close contact with humans with asymptomatic SARS-CoV-2 infection. Similar to other companion animals, horses appear to be incidental hosts because of occasional SARS-CoV-2 spillover from humans. From an epidemiological standpoint, it is important to continue to monitor the possible transmission of SARS-CoV-2 infection in equids and other domestic animals and to emphasize the risk of SARS-CoV-2 transmission from humans with clinical or asymptomatic SARS-CoV-2 infection to susceptible animals.

Author Contributions: Conceptualization, N.P., R.M.A., B.C.M., K.O.Y.L. and S.B.; data curation, N.P. and S.B.; formal analysis, S.B. and K.O.Y.L.; funding acquisition, N.P.; investigation, N.P., S.B. and K.O.Y.L.; methodology, S.B. and K.O.Y.L.; project administration, N.P. and S.B.; resources, N.P.; software, N.P., S.B. and K.O.Y.L.; supervision, N.P.; validation, S.B. and K.O.Y.L.; visualization, N.P.; writing—original draft, N.P.; writing—review and editing, N.P., R.M.A., B.C.M., S.B. and K.O.Y.L. All authors have read and agreed to the published version of the manuscript.

Funding: This research was funded by an Advancement in Equine Research Award, Boehringer Ingelheim Animal Health and the Center for Equine Health, School of Veterinary Medicine, University of California, Davis, with additional contributions from public and private donors.

Institutional Review Board Statement: The study was approved by the Institutional Animal Care and Use Committee at the University of California at Davis (protocol code 21904, approved 9 February 2020).

Data Availability Statement: Data available on request due to privacy restrictions.

Acknowledgments: The authors would like to thank all equine veterinarians who submitted samples to the laboratory.

Conflicts of Interest: The authors declare no conflict of interest.

References

1. Zheng, J. SARS-CoV-2: An emerging coronavirus that causes a global threat. *Int. J. Biol. Sci.* **2020**, *16*, 1678–1685. [CrossRef] [PubMed]
2. Damas, J.; Hughes, G.M.; Keough, K.C.; Painter, C.A.; Persky, N.S.; Corbo, M.; Hiller, M.; Koepfli, K.P.; Pfenning, A.R.; Zhao, H.; et al. Broad host range of SARS-CoV-2 predicted by comparative and structural analysis of ACE2 in vertebrates. *Proc. Natl. Acad. Sci. USA* **2020**, *117*, 22311–22322. [CrossRef]
3. Lan, J.; Ge, J.; Yu, J.; Shan, S.; Zhou, H.; Fan, S.; Zhang, Q.; Shi, X.; Wang, Q.; Zhang, L.; et al. Structure of the SARS-CoV-2 spike receptor-binding domain bound to the ACE2 receptor. *Nature* **2020**, *581*, 215–220. [CrossRef] [PubMed]
4. Lu, G.; Wang, Q.; Gao, G.F. Bat-to-human: Spike features determining 'host jump' of coronaviruses SARS-CoV, MERS-CoV, and beyond. *Trends Microbiol.* **2015**, *23*, 468–478. [CrossRef] [PubMed]
5. Abdel-Moneim, A.S.; Abdelwhab, E.M. Evidence for SARS-CoV-2 infection of animal hosts. *Pathogens* **2020**, *9*, 529. [CrossRef]
6. Hossain, M.G.; Javed, A.; Akter, S.; Saha, S. SARS-CoV-2 host diversity: An update of natural infections and experimental evidence. *J. Microbiol. Immunol. Infect.* **2021**, *54*, 175–181. [CrossRef] [PubMed]
7. Jo, W.K.; de Oliveira-Filho, E.F.; Rasche, A.; Greenwood, A.D.; Osterrieder, K.; Drexler, J.F. Potential zoonotic sources of SARS-CoV-2 infections. *Transbound. Emerg. Dis.* **2021**, *68*, 1824–1834. [CrossRef] [PubMed]
8. Bouricha, E.M.; Hakmi, M.; Akachar, J.; Belyamani, L.; Ibrahimi, A. In silico analysis of ACE2 orthologues to predict animal host range with high susceptibility to SARS-CoV-2. *3 Biotech* **2020**, *10*, 483. [CrossRef] [PubMed]
9. Buonocore, M.; Marino, C.; Grimaldi, M.; Santoro, A.; Firoznezhad, M.; Paciello, O.; Prisco, F.; D'Ursi, A.M. New putative animal reservoirs of SARS-CoV-2 in Italian fauna: A bioinformatic approach. *Heliyon* **2020**, *6*, e05430. [CrossRef]
10. Pusterla, N.; James, K.; Barnum, S.; Delwart, E. Investigation of three newly identified equine parvoviruses in blood and nasal fluid samples of clinically healthy horses and horses with acute onset of respiratory disease. *Animals* **2021**, *11*, 3006. [CrossRef] [PubMed]

11. Kooijman, L.J.; Mapes, S.M.; Pusterla, N. Development of an equine coronavirus-specific enzyme-linked immunosorbent assay to determine serologic responses in naturally infected horses. *J. Vet. Diagn. Investig.* **2016**, *28*, 414–418. [CrossRef] [PubMed]
12. Pusterla, N.; Kass, P.H.; Mapes, S.; Johnson, C.; Barnett, D.C.; Vaala, W.; Gutierrez, C.; McDaniel, R.; Whitehead, B.; Manning, J. Surveillance programme for important equine infectious respiratory pathogens in the USA. *Vet. Rec.* **2011**, *169*, 12. [CrossRef] [PubMed]
13. Pusterla, N.; Mapes, S.; Wademan, C.; White, A.; Hodzic, E. Investigation of the role of lesser characterised respiratory viruses associated with upper respiratory tract infections in horses. *Vet. Rec.* **2013**, *172*, 315. [CrossRef] [PubMed]
14. Zhao, S.; Schuurman, N.; Li, W.; Wang, C.; Smit, L.A.M.; Broens, E.M.; Wagenaar, J.A.; van Kuppeveld, F.J.M.; Bosch, B.J.; Egberink, H. Serologic screening of severe acute respiratory syndrome coronavirus 2 infection in cats and dogs during first coronavirus disease wave, the Netherlands. *Emerg. Infect. Dis.* **2021**, *27*, 1362–1370. [CrossRef] [PubMed]
15. Okba, N.M.A.; Müller, M.A.; Li, W.; Wang, C.; GeurtsvanKessel, C.H.; Corman, V.M.; Lamers, M.M.; Sikkema, R.S.; de Bruin, E.; Chandler, F.D.; et al. Severe acute respiratory syndrome coronavirus 2-specific antibody responses in coronavirus disease patients. *Emerg. Infect. Dis.* **2020**, *26*, 1478–1488. [CrossRef]
16. Dróżdż, M.; Krzyżek, P.; Dudek, B.; Makuch, S.; Janczura, A.; Paluch, E. Current state of knowledge about role of pets in zoonotic transmission of SARS-CoV-2. *Viruses* **2021**, *13*, 1149. [CrossRef]
17. Rashedi, J.; Mahdavi Poor, B.; Asgharzadeh, V.; Pourostadi, M.; Samadi Kafil, H.; Vegari, A.; Tayebi-Khosroshahi, H.; Asgharzadeh, M. Risk factors for COVID-19. *Infez. Med.* **2020**, *28*, 469–474.
18. Saba, L.; Gerosa, C.; Fanni, D.; Marongiu, F.; La Nasa, G.; Caocci, G.; Barcellona, D.; Balestrieri, A.; Coghe, F.; Orru, G.; et al. Molecular pathways triggered by COVID-19 in different organs: ACE2 receptor-expressing cells under attack? A review. *Eur. Rev. Med. Pharmacol. Sci.* **2020**, *24*, 12609–12622.
19. Deng, J.; Jin, Y.; Liu, Y.; Sun, J.; Hao, L.; Bai, J.; Huang, T.; Lin, D.; Jin, Y.; Tian, K. Serological survey of SARS-CoV-2 for experimental, domestic, companion and wild animals excludes intermediate hosts of 35 different species of animals. *Transbound. Emerg. Dis.* **2020**, *67*, 1745–1749. [CrossRef]
20. Cerino, P.; Buonerba, C.; Brambilla, G.; Atripaldi, L.; Tafuro, M.; Concilio, D.D.; Vassallo, L.; Conte, G.L.; Cuomo, M.C.; Maiello, I.; et al. No detection of SARS-CoV-2 in animals exposed to infected keepers: Results of a COVID-19 surveillance program. *Future Sci. OA* **2021**, *7*, FSO711. [CrossRef]
21. Bosco-Lauth, A.M.; Walker, A.; Guilbert, L.; Porter, S.; Hartwig, A.; McVicker, E.; Bielefeldt-Ohmann, H.; Bowen, R.A. Non susceptibility of livestock to SARS-CoV-2 infection. *Emerg. Microbes Infect.* **2021**, *10*, 2199–2201. [CrossRef] [PubMed]
22. Sailleau, C.; Dumarest, M.; Vanhomwegen, J.; Delaplace, M.; Caro, V.; Kwasiborski, A.; Hourdel, V.; Chevaillier, P.; Barbarino, A.; Comtet, L.; et al. First detection and genome sequencing of SARS-CoV-2 in an infected cat in France. *Transbound. Emerg. Dis.* **2020**, *67*, 2324–2328. [CrossRef] [PubMed]
23. Barrs, V.R.; Peiris, M.; Tam, K.W.S.; Law, P.Y.T.; Brackman, C.J.; To, E.M.W.; Yu, V.Y.T.; Chu, D.K.W.; Perera, R.A.P.M.; Sit, T.H.C SARS-CoV-2 in quarantined domestic cats from COVID-19 households or close contacts, Hong Kong, China. *Emerg. Infect. Dis* **2020**, *26*, 3071–3074. [CrossRef] [PubMed]
24. Patterson, E.I.; Elia, G.; Grassi, A.; Giordano, A.; Desario, C.; Medardo, M.; Smith, S.L.; Anderson, E.R.; Prince, T.; Patterson, G.T.; et al. Evidence of exposure to SARS-CoV-2 in cats and dogs from households in Italy. *Nat. Commun.* **2020**, *11*, 6231. [CrossRef]
25. Calvet, G.A.; Pereira, S.A.; Ogrzewalska, M.; Pauvolid-Corrêa, A.; Resende, P.C.; Tassinari, W.S.; Costa, A.P.; Keidel, L.O.; da Rocha, A.S.B.; da Silva, M.F.B.; et al. Investigation of SARS-CoV-2 infection in dogs and cats of humans diagnosed with COVID-19 in Rio de Janeiro, Brazil. *PLoS ONE* **2021**, *16*, e0250853. [CrossRef] [PubMed]
26. Ruiz-Arrondo, I.; Portillo, A.; Palomar, A.M.; Santibáñez, S.; Santibáñez, P.; Cervera, C.; Oteo, J.A. Detection of SARS-CoV-2 in pets living with COVID-19 owners diagnosed during the COVID-19 lockdown in Spain: A case of an asymptomatic cat with SARS-CoV-2 in Europe. *Transbound. Emerg. Dis.* **2021**, *68*, 973–976. [CrossRef] [PubMed]
27. Hamer, S A.; Pauvolid-Corrêa, A.; Zecca, I.B.; Davila, E.; Auckland, L.D.; Roundy, C.M.; Tang, W.; Torchetti, M.K.; Killian, M.L.; Jenkins-Moore, M.; et al. SARS-CoV-2 infections and viral isolations among serially tested cats and dogs in households with infected owners in Texas, USA. *Viruses* **2021**, *13*, 938. [CrossRef] [PubMed]
28. Venter, M.; Richter, K. Towards effective diagnostic assays for COVID-19: A review. *J. Clin. Pathol.* **2020**, *73*, 370–377. [CrossRef] [PubMed]
29. Wernike, K.; Aebischer, A.; Michelitsch, A.; Hoffmann, D.; Freuling, C.; Balkema-Buschmann, A.; Graaf, A.; Müller, T.; Osterrieder, N.; Rissmann, M.; et al. Multi-species ELISA for the detection of antibodies against SARS-CoV-2 in animals. *Transbound. Emerg. Dis.* **2021**, *68*, 1779–1785. [CrossRef] [PubMed]
30. Zhang, Q.; Zhang, H.; Gao, J.; Huang, K.; Yang, Y.; Hui, X.; He, X.; Li, C.; Gong, W.; Zhang, Y.; et al. A serological survey of SARS-CoV-2 in cat in Wuhan. *Emerg. Microbes Infect.* **2020**, *9*, 2013–2019. [CrossRef] [PubMed]
31. Fritz, M.; Rosolen, B.; Krafft, E.; Becquart, P.; Elguero, E.; Vratskikh, O.; Denolly, S.; Boson, B.; Vanhomwegen, J.; Gouilh, M.A.; et al. High prevalence of SARS-CoV-2 antibodies in pets from COVID-19+ households. *One Health* **2021**, *11*, 100192. [CrossRef] [PubMed]
32. Shi, J.; Wen, Z.; Zhong, G.; Yang, H.; Wang, C.; Huang, B.; Liu, R.; He, X.; Shuai, L.; Sun, Z.; et al. Susceptibility of ferrets, cats, dogs, and other domesticated animals to SARS-coronavirus 2. *Science* **2020**, *368*, 1016–1020. [CrossRef] [PubMed]
33. Sreenivasan, C.C.; Thomas, M.; Wang, D.; Li, F. Susceptibility of livestock and companion animals to COVID-19. *J. Med. Virol.* **2021**, *93*, 1351–1360. [CrossRef] [PubMed]

Article

Association of Equine Herpesvirus 5 with Mild Respiratory Disease in a Survey of EHV1, -2, -4 and -5 in 407 Australian Horses

Charles El-Hage [1,*], Zelalem Mekuria [1,2], Kemperly Dynon [1], Carol Hartley [1], Kristin McBride [3] and James Gilkerson [1]

1 Centre for Equine Infectious Diseases, Faculty of Veterinary and Agricultural Sciences, The University of Melbourne, Parkville, VIC 3010, Australia; mekuria.3@osu.edu (Z.M.); kemperly@gmail.com (K.D.); carolah@unimelb.edu.au (C.H.); jrgilk@unimelb.edu.au (J.G.)
2 Global One Health Initiative, Ohio State University, Columbus, OH 43210, USA
3 UNSW Medicine, University of New South Wales, Sydney, NSW 2052, Australia; kmcbride@kirby.unsw.edu.au
* Correspondence: cmeh@unimelb.edu.au; Tel.:+61-417166029

Citation: El-Hage, C.; Mekuria, Z.; Dynon, K.; Hartley, C.; McBride, K.; Gilkerson, J. Association of Equine Herpesvirus 5 with Mild Respiratory Disease in a Survey of EHV1, -2, -4 and -5 in 407 Australian Horses. *Animals* **2021**, *11*, 3418. https://doi.org/10.3390/ani11123418

Academic Editors: Amir Steinman and Oran Erster

Received: 7 November 2021
Accepted: 25 November 2021
Published: 30 November 2021

Simple Summary: Infectious respiratory diseases in horses represent a major health and welfare problem. Although equine influenza is well reported as a cause of respiratory disease in most continents, Australia is free of EIV despite an outbreak in two states in 2007. Horses in Victoria were tested to demonstrate proof of freedom from EIV, hence samples were able to be subsequently tested for this study with the knowledge that EIV was not present as a potential cause of any disease. The equine alphaherpesviruses, EHV1 and -4 are well known agents of equine respiratory disease. The gammaherpesviruses EHV2 and -5 on the other hand are often isolated from clinically healthy horses despite a known association in some disease processes. The consequences of infection with these enigmatic viruses remains unknown. The investigation of several hundred horses with and without respiratory disease provided valuable information in terms of association. The salient findings of this study determined that a large proportion of normal horses were positive for the gammaherpesviruses EHV2 and -5 using PCR methods. However, horses shedding EHV5 were more likely to have had signs of respiratory disease. Like EHV2, EHV5 is a gammaherpesvirus commonly found in horses: its significance is unclear, though it is closely related to the Epstein–Barr virus, the agent responsible for glandular fever in humans. These viruses are known to interfere with the immune response and have potentially wide-ranging effects on infected hosts. This study has added to our awareness of these equine herpesviruses and should stimulate further studies to determine exact causation and consequences of infection.

Abstract: Equine herpesviruses (EHVs) are common respiratory pathogens in horses; whilst the alphaherpesviruses are better understood, the clinical importance of the gammaherpesviruses remains undetermined. This study aimed to determine the prevalence of, and any association between, equine respiratory herpesviruses EHV1, -2, -4 and -5 infection in horses with and without clinical signs of respiratory disease. Nasal swabs were collected from 407 horses in Victoria and included clinically normal horses that had been screened for regulatory purposes. Samples were collected from horses during Australia's equine influenza outbreak in 2007; however, horses in Victoria required testing for proof of freedom from EIV. All horses tested in Victoria were negative for EIV, hence archived swabs were available to screen for other pathogens such as EHVs. Quantitative PCR techniques were used to detect EHVs. Of the 407 horses sampled, 249 (61%) were clinically normal, 120 (29%) presented with clinical signs consistent with mild respiratory disease and 38 (9%) horses had an unknown clinical history. Of the three horses detected shedding EHV1, and the five shedding EHV4, only one was noted to have clinical signs referable to respiratory disease. The proportion of EHV5-infected horses in the diseased group (85/120, 70.8%) was significantly greater than those not showing signs of disease (137/249, 55%). The odds of EHV5-positive horses demonstrating clinical signs of respiratory disease were twice that of EHV5-negative horses (OR 1.98, 95% CI 1.25 to 3.16). No quantitative difference between mean loads of EHV shedding between diseased and non-diseased horses was detected. The clinical significance of respiratory gammaherpesvirus infections in horses remains to be determined; however, this survey adds to the mounting body of evidence associating EHV5 with equine respiratory disease.

Keywords: gammaherpesvirus; horses; respiratory disease; equine herpesvirus 1, -2, -4, -5; equine influenza; quantitative PCR

1. Introduction

Equine herpesviruses (EHVs) are common respiratory pathogens in equids. These viruses have serious health and welfare outcomes in horses and significant financial consequences worldwide [1–4]. Both the alphaherpesviruses EHV1 and -4 are transmitted by the respiratory route, although respiratory disease is more commonly attributed to EHV4. The clinical importance of the gammaherpesviruses EHV2 and -5 is less clear [5–7]. This lack of clarity may be attributed to the frequent detection of gammaherpesviruses in horses with and without clinical signs of disease, under both experimental and field conditions [8–18]. Although outbreaks of disease caused by alphaherpesviruses are commonly reported in horses, shedding from the respiratory tract is often of short duration, and usually only detected in a minority of the population [4,19–24]. Many studies have detected gammaherpesviruses in a large percentage of horses within a population, often with few clinical signs of disease [9,14,15,25,26]. Although the gammaherpesviruses are commonly detected in clinical samples from horses, there are differences in the frequency of detection of these two viruses. The relative prevalence of these viruses varies in different studies, with some studies showing higher detection of EHV2 than EHV5 [12,15,27], and others showing EHV5 as more prevalent [17,25,26,28–30]. Several studies since 2007 have reported an association between EHV5 detection and a pulmonary fibrotic condition of horses, equine multi-nodular pulmonary fibrosis (EMPF) [18,31–34].

Individual horses can be infected with multiple herpesvirus species [29,35–40]. It has been hypothesised that infection with equine gammaherpesvirus may result in immunosuppression and, consequently, increased susceptibility to new or reactivated infections. While equine gammaherpesviruses contain many potential immunomodulating genes, [41,42] and gammaherpesvirus-mediated immunosuppression has been demonstrated in other species [43,44], this has not been reported as extensively in horses.

The opportunity to sample diseased and clinically normal horses arose during Australia's only recorded equine influenza (EI) outbreak in 2007. The outbreak was limited to states north of Victoria, which remained free of EI. Equine respiratory samples were collected in Victoria for EI exclusion. This formed a central part of the outbreak investigation to confirm that equine influenza virus (EIV) had not spread to Victoria and was required for horse movement permits.

The aim of the study was to examine the prevalence of four endemic equine respiratory herpesviruses, and to determine if there was any association between infection and clinical respiratory disease.

2. Materials and Methods

2.1. Study Population

The study population consisted of 407 horses in Victoria with and without clinical signs of respiratory disease during the Australian EI outbreak from August 2007 to January 2008. These horses were sampled for the purposes of EI exclusion if they (i) had clinical signs of respiratory disease, (ii) potentially had contact with infected horses or (iii) required movement clearances. Clinical signs of mild respiratory disease were recorded as one or more of the following signs: coughing, pyrexia (temperature >38.5 °C) and/or nasal discharge [45]. A total of 522 nasal swabs were collected from these horses for exclusion of EIV. Vaccination histories were not recorded. In Australia, EHV1 and -4 (Duvaxyn, EHV-1, 4, Zoetis P/L, Castle Hill, NSW, Australia) and *Streptococcus equi* sub-species *equi* (Equivac-S™, Zoetis P/L, Castle Hill, NSW, Australia) vaccines are commonly used [46] while EIV vaccination is not permitted in Australia, unless for export purposes.

2.2. Nasal Swabs

Nasal swabs were collected using swabs with a 15 cm wooden shaft and a cotton tip (Interpath Services, Heidelberg West, VIC, Australia 163KS01) [47]. Swab tips were placed into 5 mL of Brain Heart Infusion (BHI) broth-based viral transport medium (BHI 3.7% *w*/*v* in sterile distilled water with penicillin 5000 U/mL, gentamicin 0.1 mg/mL, streptomycin 5 mg/mL and 200 µg/mL fungizone (Sigma Healthcare, Rowville, VIC, Australia)). Following testing for EI, swabs in transport media were stored at −80 °C.

2.3. Nucleic Acid Extraction from Nasal Swabs

Nucleic acid extraction from nasal swabs was performed using an automated system (X-tractor Gene, Qiagen) using the QIAamp® Virus Biorobot 9604 Kit (QIAGEN, Germantown, MD, USA) according to the manufacturer's recommendations (https://www.qiagen.com (accessed on 20 December 2020)). Virus culture supernatants were included as known positive control samples and were also extracted using this method [48]. Viruses used as positive controls were EHV1.438/77 [49], EHV4.405/76 [49], EHV2.86/67 [50] and EHV5.2-141 [51].

2.4. Quantitative PCR Assays

All quantitative PCR (qPCR) tests were performed in a Stratagene© MxPro Mx3000P real-time PCR machine (Agilent Technologies Inc., Santa Clara, CA, USA), and analysed with the machine's software with cycle threshold values assigned using the default threshold algorithm. Standard curves were generated from cycle thresholds of samples with known virus concentrations and genome copy numbers.

EHV1 and EHV4 were detected in a multiplex Taqman assay with the primers and probes targeting the EHV1 glycoprotein H gene, and the EHV4 intergenic region between open reading frames 73 and 74 (Appendix A, Table A1). These were used in a 20 µL reaction containing Brilliant qPCR Multiplex master mix (Stratagene, Agilent Technologies Inc., Santa Clara, CA, USA), 200 nM of each forward and reverse primer and probes, 30 nM ROX reference dye and 5 µL of sample DNA. The reaction thermocycling conditions were 95 °C for 10 min, followed by 40 cycles of 95 °C for 30 s and 60 °C for 1 min. Samples with a Ct value of $\leq$35 were considered positive.

Equine herpesviruses 2 and 5 were detected in two separate qPCR assays using SYTO® 9 (Invitrogen™, Thermo Fisher Scientific, Waltham, MA, USA) as a double stranded DNA-binding dye. Equine herpesvirus 2 primers were designed to the glycoprotein B gene and equine herpesvirus 5 primers were designed to the glycoprotein H gene of EHV5 (Table A1). Each 25µL reaction volume contained 2 µg/mL SYTO9, 0.2 U GoTaq (Promega Corporation. Madison, WI, USA), 300 nM of the appropriate forward and reverse primers (Table A1) and 1.5 mM $MgCl_2$ in the GoTaq reaction buffer as recommended by the manufacturer (Promega Corporation, Madison, WI, USA). Thermocycling of reactions proceeded at 94 °C for 15 min, then 40 cycles of 94 °C for 15 s, 60 °C annealing for 30 s and 72 °C extension for 30 s. The melting curve analysis of each amplicon was analysed after one cycle of 95 °C for 1 min, 55 °C for 30 s and 95 °C for 30 s. Samples were considered positive if the melting temperature of the amplicon was within the range specified below and the cycle threshold was below 35 in the EHV2 [52], and 37 in the EHV5 [41] assays, respectively. The melting temperature of the amplicon was determined using four diverse EHV2 isolates [10,51], and three EHV5 isolates [51] and occurred within the range 79 to 81 °C for EHV2 and 80 to 82 °C for EHV5. Positive control viruses EHV2.86/67 and EHV5.2-141 and nuclease-free water were included for each 96-well extraction and PCR plate.

2.5. Statistical Analysis

Comparisons of two proportions were determined by Fisher's exact test. A two-sample t-test was used to compare mean quantitative cycles between samples from non-diseased and diseased horses. Any two-sided Student's t-test with a p value less than 0.05 was considered to be significant. Logistic regression methods were utilised to test for interaction. Statistical analysis was performed using Stata 12.1 Windows software (StataCorp, College Station, TX, USA).

3. Results

3.1. Stratification of Horses in Terms of Respiratory Disease

Of the 407 horses sampled, 249 (61%) were clinically normal, 120 (29%) presented with clinical signs consistent with mild respiratory disease and 38 (9%) horses had an unknown clinical history (Table 1). For instances where multiple samples were taken at a single time point, a horse was reported as infected if viral DNA was detected in any sample collected at that time.

Table 1. Clinical status of horses and detection of. Equine Herpesvirus 1, -4, -2 and -5 from nasal swabs.

Virus Detected	Respiratory Disease Signs			
	Negative	Positive	Not Recorded	Total
EHV-1 only	1	0	0	1
EHV-2 only	29	4	0	33
EHV-5 only	105	70	22	197
EHV-1 and EHV-4	2	0	0	2
EHV-2 and EHV-5	31	14	4	49
EHV-4 and EHV-5	0	1	1	2
EHV-2, EHV-5 and EHV-4	1	0	0	1
No detection	80	31	11	122
Total	249	120	38	407

3.2. Equine Herpesvirus Infections

3.2.1. Equine Herpesvirus 1

Equine herpesvirus 1 was detected in three horses (Table 1). The viral loads detected in these samples ranged from $10^{6.45}$, $10^{7.91}$ and $10^{9.22}$ genome copies/mL of nasal swab. None of these horses exhibited any clinical signs of respiratory disease at the time of sampling.

3.2.2. Equine Herpesvirus 4

Five horses were EHV4 positive. Three of these horses were clinically normal when sampled. The highest EHV4 load was $10^{8.45}$ genome copies/mL nasal swab from a horse of unknown clinical status. There were insufficient data for a meaningful comparison of aphaherpesvirus shedding between diseased and normal horses.

3.3. Equine Gammaherpesvirus Infections, EHV2 and -5

In total, 83 (20.4%) of the 407 horses sampled were EHV2 positive and 249 horses (61.2%) were positive by qPCR for EHV5 (Table 1). There were no differences between the mean viral load of EHV2 or EHV5 detected in diseased and non-diseased horses (Figure 1) There was, however, a statistically significant difference ($p = 0.004$) between the proportion of horses in the diseased group shedding EHV5 (85/120, 70.8%) compared to the proportion of horses in the non-diseased group that were shedding EHV5 at the time of sampling (137/249, 55%). The odds of respiratory disease in EHV5-positive horses were twice that of EHV5-negative horses (OR 1.98, 95% CI 1.25 to 3.16). The proportion of horses with detectable EHV2 was significantly higher in non-diseased horses (61/249, 24.5%) compared to the diseased group (18/120, 15.0%) ($p = 0.042$). The odds of EHV2-positive horses also exhibiting clinical signs of disease were approximately half that of EHV2-negative horses (OR 0.54, 95% CI 0.30 to 0.97).

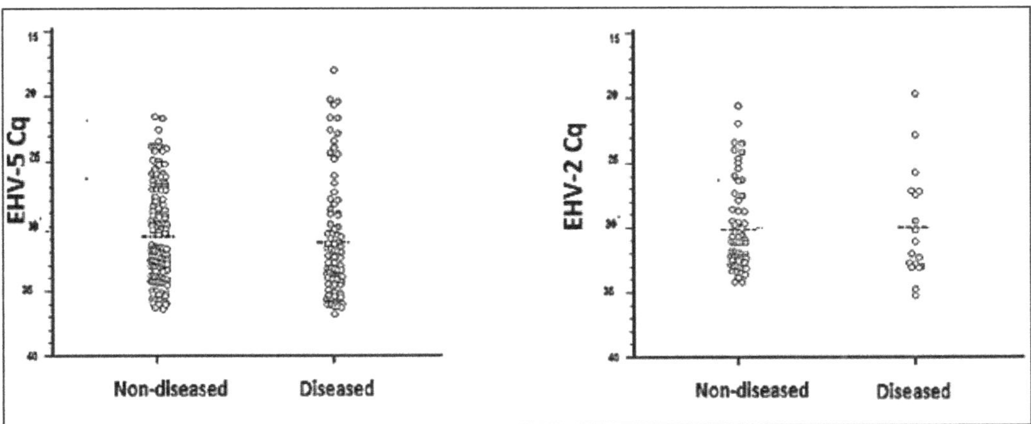

Figure 1. Quantification cycles (Cq) values considered positive for Equine herpesvirus 2 and -5 in nasal swabs of horses with and without clinical signs of disease (diseased and non-diseased). The horizontal line indicates the mean Cq for each group, none of which were statistically different between groups.

3.4. Concurrent Equine Herpesvirus Infections

Of the 407 horses sampled in this survey, 54 (13.3%) were shedding multiple equine herpesviruses (Table 1). Two of the three horses shedding detectable levels of EHV1 were concurrently shedding EHV4. Three of the five EHV4-positive horses were also shedding EHV5, and a fourth was shedding EHV2. One horse that was clinically normal was shedding EHV2, -4 and -5 concurrently. The horse shedding the highest EHV4 load of $10^{8.45}$ copies/mL nasal swab was also shedding $10^{7.94}$ copies/mL of EHV5; however, the disease status of this horse was unknown.

Fifty of the eighty three horses (60.2%) shedding EHV2 were also shedding EHV5; however, there was no greater likelihood of EHV5 detection in these horses compared to those without detectable EHV2 (199/324, 61.4%; $p = 0.90$). In addition, these co-infected horses were no more likely to exhibit signs of disease (14/46, 30.4%) than those shedding only EHV2 (4/33, 12.1%; $p = 0.063$) or EHV5 (71/176, 40.3%; $p = 0.24$). Logistic regression showed no correlation between dual EHV2 and -5 shedding and clinical signs of respiratory disease ($p = 0.41$). Hence, the association of EHV5 infection and increased likelihood of disease was not modified by the presence or absence of EHV2 infection.

4. Discussion

Equine herpesvirus infections were commonly detected in samples from the respiratory tract, irrespective of clinical disease status at the time of sampling. Approximately 40% of horses were shedding at least one herpesvirus at the time of sampling (Table 1). In total, 67.9% of horses with no obvious clinical disease were shedding detectable levels of at least one herpesvirus. Detection of the alphaherpesviruses in a small proportion of horses (2%, $n = 8$) contrasted markedly with the high frequency of shedding of the equine gammaherpesviruses (69.3%, $n = 282$). Although many clinically normal horses were infected, a significantly high proportion of horses with clinical signs of respiratory disease were shedding EHV5. No such association was detected in horses infected with EHV1, -4 and -2.

The increased proportion of horses shedding EHV5 among diseased horses in this study may reflect the contribution of EHV5 to respiratory disease. Alternatively, this shedding may have been reactivated as a consequence of a respiratory disease-associated inflammatory response. The spectrum of clinical disease (or lack of) following gammaherpesvirus infections in horses may be due to a range of factors including virus strain and load, host factors such as age [11,53], and immune responses [54]. Each of these complex factors has been explored in several studies and may help to explain the lack of disease seen in many infected horses. EHV5 is persistently associated with EMPF while it is also regularly detected in both clinically normal and diseased horses [11,13–16,25,27,31–33]. This study showed a significant difference in the proportion of horses shedding EHV5 in the diseased group, such that the odds of disease signs in EHV5-positive horses were twice that of EHV5-negative horses. This difference may be the result of lytic EHV5 infection causing the clinical signs, or that EHV5 is reactivated by infection/inflammation by another agent. B-lymphocytes are a latent reservoir for EHV2 and EHV5, and other sites may exist which have not yet been identified [44,55–57]. However, simple reactivation of shedding via B-lymphocytes recruited to these sites does not account for the difference in the clinical associations of EHV5 and EHV2 in this study. Other studies have also shown a protective effect of EHV2 against *Rhodococcus equi* infection [58]. Whether EHV2 and EHV5 each occupy distinct niches within the respiratory tract, or whether each recruit different types of inflammatory cells that might be protective or immunopathogenic, remains unknown.

The higher incidence of EHV2 in non-diseased horses in this study is consistent with those of previous studies and continues to confound our understanding of the role of this virus, if any, in equine respiratory tract disease. The prevalence of EHV2 infection in large numbers of clinically normal horses has been widely reported [13–16,26,27]; however, several studies have identified associations between EHV2 infection and mild respiratory disease, particularly in foals [11,13,14,17,59,60].

Quantification of gammaherpesvirus shedding may enable an association to be made between viral load and clinical disease. In humans, an age-range-specific correlation exists between the levels of the gammaherpesvirus Epstein–Barr virus (EBV) in blood, and the presence of clinical disease [61]; however, there is currently little evidence in this or other studies to support an association with acute respiratory disease and gammaherpesvirus load in horses [11,58,62]. Multiple factors are likely to be required for gammaherpesvirus-mediated disease in horses, rather than solely lytic infections. Alternatively, nasal samples may not be the most appropriate samples as predictors for clinical respiratory disease. This is supported by a recent publication linking high viral loads of EHV5 in bronchoalveolar lavage fluid to EMPF [18].

The detection of alphaherpesviruses is reported in a minority of horses within most populations [17,19,20,63–66]. Five horses without signs of disease were shedding high levels of either alphaherpesvirus EHV1 or -4, consistent with the "cycle of silent herpesvirus shedding" and spread [63,66,67]. The reactivation of latent alphaherpesvirus infection is associated with subclinical viral shedding [20] and can occur following stressful events such as social re-grouping, weaning and long-distance transport [35,65,68,69]. Despite these factors, the levels of detection of equine herpesviruses in this study population were consistent with other studies that have reported ranges of 0–10% for the alphaherpesviruses and 0–100% for the gammaherpesviruses [12,14,16,20,51,63,65,67,70]. The reactivation of latent herpesviruses following a single immunosuppressive event may explain the detection of multiple herpesviruses. This phenomenon has been documented in humans with prolonged sepsis [71]. Shedding of multiple EHVs was detected in 14% (57/407) of horses. Four of the six horses (67%) infected with the alphaherpesviruses EHV1 and -4 were infected by either another alpha- or a gammaherpesvirus(es).

Although Victoria remained free of EIV during Australia's only recorded EI outbreak, field staff faced logistical challenges and were often time poor. However, samples were successfully collected, and testing was not compromised, ensuring that EIV could be ruled out in all samples analysed. The lack of comprehensive histories and clinical detail for every horse including age, vaccination status and time-course of clinical disease may have limited the analysis of data. The inclusion of the 38 horses of unknown clinical status was made to assist the determination of overall prevalence. A separate analysis of these horses did not show any statistical difference in the proportion of EHV infection in these horses compared with those of known status (diseased and non-diseased).

5. Conclusions

The clinical significance of respiratory gammaherpesvirus infections in horses remains to be determined; however, this survey adds to the mounting body of evidence associating EHV5 with equine respiratory disease. The task of identifying a definitive role of the equine gammaherpesviruses as the cause of respiratory disease on a case-by-case basis remains challenging, since the precise role of both EHV2 and -5 and their relation to clinical disease is likely to be complex and remains to be elucidated for these enigmatic viruses.

Author Contributions: Conceptualisation, J.G., C.H. and C.E.-H.; methodology, C.H., K.D., K.M. and J.G.; validation, C.H., K.D., Z.M., K.M. and J.G.; formal analysis, C.E.-H., C.H., K.D., Z.M. and J.G.; investigation, C.E.-H., C.H., Z.M. and J.G.; writing—C.E.-H., C.H. and J.G.; writing—C.H., K.D., Z.M. and K.M.; supervision, J.G. and C.H.; project administration, J.G. and C.E.-H.; All authors have read and agreed to the published version of the manuscript.

Funding: This research received no external funding.

Institutional Review Board Statement: Ethical review and approval were waived for this study, due to access of archived material collected for surveillance and demonstration of freedom from equine influenza virus.

Informed Consent Statement: Not applicable.

Data Availability Statement: Not applicable.

Acknowledgments: The authors wish to thank the Chief Veterinary Officer's unit and staff within the (formerly named) Department of Environment and Primary Industries Victoria and Doctors Mark Hawes and Simone Warner for their valued assistance. Thanks to Nino Ficorilli and Cynthia Brown for their valued technical assistance and Garry Anderson who provided skilled statistical advice and assistance.

Conflicts of Interest: The authors declare no conflict of interest.

Appendix A

Table A1. Primers used in this study.

Primer Name	Primer Sequence 5′ to 3′
EHV1.gH.F	GCC CGA CAC CTA CAT AAC C
EHV1.gH.R	GGC ATA AAA CCA CAC CAA CC
EHV1.gH.Probe	FAM-GCG ACC ACA AAA AGC AAC CC-BHQ1
EHV4.ORF73/74.F	GGC AAC CTA CCC GAA GAT G
EHV4.ORF73/74.R	CAA CAA CCA CCA GCA ACA A
EHV4.ORF73/74.Probe	CAL Fluor Orange 560-CCC CCA AAC CGC AAA CCA CT-BHQ1
EHV2.gB.1822.F	ACC CTC AAC CTG ACT GAC AT
EHV2.gB.1953.R	TCA AAC ACG TTG GAC AGC CT
EHV5.gH.F	TGT GTG CAA TGT TTC TGG GGG
EHV5.gH.R	CGC TGC CCA ACA CGT CCC TT

References

1. Slater, J. Equine herpesviruses. In *Equine Infectious Diseases*; Sellon, D., Long, M., Eds.; Elsevier: Amsterdam, The Netherlands, 2014; pp. 151–168.
2. Matsumura, T.; Sugiura, T.; Imagawa, H.; Fukunaga, Y.; Kamada, M. Epizootiological Aspects of Type 1 and Type 4 Equine Herpesvirus Infections among Horse Populations. *J. Vet. Med. Sci.* **1992**, *54*, 207–211. [CrossRef] [PubMed]

3. Studdert, M. (Ed.) *Virus Infections of Vertebrates, Virus Infections of Equines*; Elsevier Science Publishers: Amsterdam, The Netherlands, 1996; pp. 281–284.
4. Allen, G.P.; Kydd, J.H.; Slater, J.D.; Smith, K.C. Equid herpesvirus-1 (EHV-1) and -4 (EHV-4) infections. In *Infectious Diseases of Livestock*; Tustin, J.C.A.R., Ed.; Oxford Press: Cape Town, South Africa, 2004; pp. 829–859.
5. Allen, G.P.; Murray, M. Equid herpesvirus 2 and equid herpesvirus 5 infections. In *Infectious Diseases of Livestock*; Tustin, J.C.A.R., Ed.; Oxford Press: Cape Town, South Africa, 2004; pp. 860–867.
6. Hartley, C.A.; Dynon, K.J.; Mekuria, Z.H.; El-Hage, C.M.; Holloway, S.A.; Gilkerson, J.R. Equine gammaherpesviruses: Perfect parasites? *Vet. Microbiol.* **2013**, *167*, 86–92. [CrossRef]
7. Fortier, G.; van Erck, E.; Pronost, S.; Lekeux, P.; Thiry, E. Equine gammaherpesviruses: Pathogenesis, epidemiology and diagnosis. *Vet. J.* **2010**, *186*, 148–156. [CrossRef] [PubMed]
8. Borchers, K.; Lieckfeldt, D.; Ludwig, A.; Fukushi, H.; Allen, G.; Fyumagwa, R.; Hoare, R. Detection of Equid herpesvirus 9 DNA in the trigeminal ganglia of a Burchell's zebra from the Serengeti ecosystem. *J. Vet. Med. Sci.* **2008**, *70*, 1377–1381. [CrossRef]
9. Borchers, K.; Wolfinger, U.; Goltz, M.; Broll, H.; Ludwig, H. Distribution and relevance of equine herpesvirus type 2(EHV-2) infections. *Arch. Virol.* **1997**, *142*, 917–928. [CrossRef] [PubMed]
10. Browning, G.F.; Studdert, M.J. Epidemiology of equine herpesvirus 2 (equine cytomegalovirus). *J. Clin. Microbiol.* **1987**, *25*, 13–16. [CrossRef]
11. Dunowska, M.; Howe, L.; Hanlon, D.; Stevenson, M. Kinetics of Equid herpesvirus type 2 infections in a group of Thoroughbred foals. *Vet. Microbiol.* **2011**, *152*, 176–180. [CrossRef]
12. Dunowska, M.; Wilks, C.R.; Studdert, M.; Meers, J. Equine respiratory viruses in foals in New Zealand. *N. Z. Vet. J.* **2002**, *50*, 140–147. [CrossRef]
13. Dunowska, M.; Wilks, C.; Studdert, M.; Meers, J. Viruses associated with outbreaks of equine respiratory disease in New Zealand. *N. Z. Vet. J.* **2002**, *50*, 132–139. [CrossRef]
14. Fortier, G.; van Erck, E.; Fortier, C.; Richard, E.; Pottier, D.; Pronost, S.; Miszczak, F.; Thiry, E.; Lekeux, P. Herpesviruses in respiratory liquids of horses: Putative implication in airway inflammation and association with cytological features. *Vet. Microbiol.* **2009**, *139*, 34–41. [CrossRef]
15. Torfason, E.G.; Thorsteinsdottir, L.; Torsteinsdóttir, S.; Svansson, V. Study of equid herpesviruses 2 and 5 in Iceland with a type-specific polymerase chain reaction. *Res. Vet. Sci.* **2008**, *85*, 605–611. [CrossRef]
16. Bell, S.A.; Balasuriya, U.B.; Gardner, I.A.; Barry, P.A.; Wilson, W.D.; Ferraro, G.L.; MacLachlan, N.J. Temporal detection of equine herpesvirus infections of a cohort of mares and their foals. *Vet. Microbiol.* **2006**, *116*, 249–257. [CrossRef] [PubMed]
17. Wang, L.; Raidal, S.L.; Pizzirani, A.; Wilcox, G.E. Detection of respiratory herpesviruses in foals and adult horses determined by nested multiplex PCR. *Vet. Microbiol.* **2007**, *121*, 18–28. [CrossRef] [PubMed]
18. Pusterla, N.; Magdesian, K.G.; Mapes, S.M.; Zavodovskaya, R.; Kass, P.H. Assessment of quantitative polymerase chain reaction for equine herpesvirus-5 in blood, nasal secretions and bronchoalveolar lavage fluid for the laboratory diagnosis of equine multinodular pulmonary fibrosis. *Equine Vet. J.* **2016**, *49*, 34–38. [CrossRef] [PubMed]
19. Brown, J.A.; Mapes, S.; Ball, B.A.; Hodder, A.D.J.; Liu, I.K.M.; Pusterla, N. Prevalence of equine herpesvirus-1 infection among Thoroughbreds residing on a farm on which the virus was endemic. *J. Am. Vet. Med. Assoc.* **2007**, *231*, 577–580. [CrossRef]
20. Pusterla, N.; Mapes, S.; Madigan, J.E.; MacLachlan, N.J.; Ferraro, G.L.; Watson, J.L.; Spier, S.J.; Wilson, W.D. Prevalence of EHV-1 in adult horses transported over long distances. *Vet. Rec.* **2009**, *165*, 473–475. [CrossRef]
21. Gilkerson, J.; Jorm, L.R.; Love, D.N.; Lawrence, G.L.; Whalley, J.M. Epidemiological investigation of equid herpesvirus-4 (EHV-4) excretion assessed by nasal swabs taken from thoroughbred foals. *Vet. Microbiol.* **1994**, *39*, 275–283. [CrossRef]
22. Gilkerson, J.; Whalley, J.; Drummer, H.; Studdert, M.; Love, D. Epidemiology of EHV-1 and EHV-4 in the mare and foal populations on a Hunter Valley stud farm: Are mares the source of EHV-1 for unweaned foals. *Vet. Microbiol.* **1999**, *68*, 27–34. [CrossRef]
23. Gibson, J.S.; Slater, J.D.; Awan, A.R.; Field, H.J. Pathogenesis of equine herpesvirus-1 in specific pathogen-free foals: Primary and secondary infections and reactivation. *Arch. Virol.* **1992**, *123*, 351–366. [CrossRef]
24. Yactor, J.; Lunn, K.F.; Morley, P.; Barnett, C.D.; Kohler, A.K.; Kasper, K.S.; Kivi, A.J.; Lunn, D.P. Detection of nasal shedding of EHV-1 and 4 at equine show events and sales by multiplex real-time PCR. In Proceedings of the American Association of Equine Practitioners (AAEP), San Antonio, TX, USA, 6 December 2006; pp. 223–227.
25. Wang, L. *An Investigation of the Association between Herpes Viruses and Respiratory Disease*; Murdoch University: Perth, WA, Australia, 2003.
26. Back, H.; Ullman, K.; Berndtsson, L.T.; Riihimäki, M.; Penell, J.; Ståhl, K.; Valarcher, J.-F.; Pringle, J. Viral load of equine herpesviruses 2 and 5 in nasal swabs of actively racing Standardbred trotters: Temporal relationship of shedding to clinical findings and poor performance. *Vet. Microbiol.* **2015**, *179*, 142–148. [CrossRef]
27. Nordengrahn, A.; Merza, M.; Ros, C.; Lindholm, A.; Hannant, D. Prevalence of equine herpesvirus types 2 and 5 in horse populations by using type-specific PCR assays. *Vet. Res.* **2002**, *33*, 251–259. [CrossRef] [PubMed]
28. Diallo, I.S.; Hewitson, G.R.; De Jong, A.; Kelly, M.A.; Wright, D.J.; Corney, B.G.; Rodwell, B.J. Equine herpesvirus infections in yearlings in South-East Queensland. *Arch. Virol.* **2008**, *153*, 1643–1649. [CrossRef]
29. McBrearty, K.A.; Murray, A.; Dunowska, M. A survey of respiratory viruses in New Zealand horses. *N. Z. Vet. J.* **2013**, *61*, 254–261. [CrossRef]

30. Hue, E.S.; Fortier, G.D.; Fortier, C.I.; Leon, A.M.; Richard, E.A.; Legrand, L.J.; Pronost, S.L. Detection and quantitation of equid gammaherpesviruses (EHV-2, EHV-5) in nasal swabs using an accredited standardised quantitative PCR method. *J. Virol. Methods* **2014**, *198*, 18–25. [CrossRef] [PubMed]
31. Back, H.; Kendall, A.; Grandón, R.; Ullman, K.; Treiberg-Berndtsson, L.; Ståhl, K.; Pringle, J. Equine Multinodular Pulmonary Fibrosis in association with asinine herpesvirus type 5 and equine herpesvirus type 5: A case report. *Acta Vet. Scand.* **2012**, *54*, 57. [CrossRef] [PubMed]
32. Williams, K.J.; Maes, R.; Del Piero, F.; Lim, A.; Wise, A.; Bolin, D.C.; Caswell, J.; Jackson, C.; Robinson, N.E.; Derksen, F.; et al. Equine Multinodular Pulmonary Fibrosis: A Newly Recognized Herpesvirus-Associated Fibrotic Lung Disease. *Vet. Pathol.* **2007**, *44*, 849–862. [CrossRef]
33. Marenzoni, M.L.; Passamonti, F.; Lepri, E.; Cercone, M.; Capomaccio, S.; Cappelli, K.; Felicetti, M.; Coppola, G.; Coletti, M.; Thiry, E. Quantification of Equid herpesvirus 5 DNA in clinical and necropsy specimens collected from a horse with equine multinodular pulmonary fibrosis. *J. Vet. Diagn. Investig.* **2011**, *23*, 802–806. [CrossRef]
34. Williams, K.J.; Robinson, N.E.; Lim, A.; Brandenberger, C.; Maes, R.; Behan, A.; Bolin, S.R. Experimental Induction of Pulmonary Fibrosis in Horses with the Gammaherpesvirus Equine Herpesvirus 5. *PLoS ONE* **2013**, *8*, e77754. [CrossRef] [PubMed]
35. Edington, N.; Welch, H.; Griffiths, L. The prevalence of latent Equid herpesviruses in the tissues of 40 abattoir horses. *Equine Vet. J.* **1994**, *26*, 140–142. [CrossRef]
36. Purewal, A.S.; Smallwood, A.V.; Kaushal, A.; Adegboye, D.; Edington, N. Identification and control of the cis-acting elements of the immediate early gene of equid herpesvirus type 1. *J. Gen. Virol.* **1992**, *73*, 513–519. [CrossRef]
37. Welch, H.M.; Bridges, C.G.; Lyon, A.M.; Griffiths, L.; Edington, N. Latent equid herpesviruses 1 and 4: Detection and distinction using the polymerase chain reaction and co-cultivation from lymphoid tissues. *J. Gen. Virol.* **1992**, *73*, 261–268. [CrossRef] [PubMed]
38. Marenzoni, M.L.; Bietta, A.; Lepri, E.; Proietti, P.C.; Cordioli, P.; Canelli, E.; Stefanetti, V.; Coletti, M.; Timoney, P.J.; Passamonti, F. Role of equine herpesviruses as co-infecting agents in cases of abortion, placental disease and neonatal foal mortality. *Vet. Res. Commun.* **2013**, *37*, 311–317. [CrossRef]
39. Pusterla, N.; Mapes, S.; Wademan, C.; White, A.; Hodzic, E. Investigation of the role of lesser characterised respiratory viruses associated with upper respiratory tract infections in horses. *Vet. Rec.* **2013**, *172*, 315. [CrossRef] [PubMed]
40. Dynon, K.; Black, W.; Ficorilli, N.; Hartley, C.; Studdert, M. Detection of viruses in nasal swab samples from horses with acute, febrile, respiratory disease using virus isolation, polymerase chain reaction and serology. *Aust. Vet. J.* **2007**, *85*, 46–50. [CrossRef] [PubMed]
41. Mekuria, Z.H. Equine herpesvirus 5; Cellular and molecular biology. In *Veterinary Science*; University of Melbourne: Melbourne, VIC, Australia, 2015.
42. Telford, E.A.; Watson, M.S.; Aird, H.C.; Perry, J.; Davison, A.J. The DNA Sequence of Equine Herpesvirus 2. *J. Mol. Biol.* **1995**, *249*, 520–528. [CrossRef] [PubMed]
43. Silins, S.L.; Sherritt, M.A.; Silleri, J.M.; Cross, S.M.; Elliott, S.L.; Bharadwaj, M.; Le, T.T.T.; Morrison, L.E.; Khanna, R.; Moss, D.J.; et al. Asymptomatic primary Epstein-Barr virus infection occurs in the absence of blood T-cell repertoire perturbations despite high levels of systemic viral load. *Blood* **2001**, *98*, 3739–3744. [CrossRef]
44. Sitki-Green, D.L.; Edwards, R.H.; Covington, M.M.; Raab-Traub, N. Biology of Epstein-Barr Virus during Infectious Mononucleosis. *J. Infect. Dis.* **2004**, *189*, 483–492. [CrossRef]
45. Anonymous. Information for Victorian Veterinarians. Agriculture Victoria. 2007. Available online: https://agriculture.vic.gov.au/biosecurity/animal-diseases/horse-diseases/equine-influenza (accessed on 20 May 2020).
46. Goyen, K.A.; Wright, J.D.; Cunneen, A.; Henning, J. Playing with fire—What is influencing horse owners' decisions to not vaccinate their horses against deadly Hendra virus infection? *PLoS ONE* **2017**, *12*, e0180062. [CrossRef]
47. Kirkland, P.; Davis, R.; Wong, D.; Ryan, D.; Hart, K.; Corney, B.; Hewitson, G.; Cooper, K.; Biddle, A.; Eastwood, S.; et al. The first five days: Field and laboratory investigations during the early stages of the equine influenza outbreak in Australia, 2007. *Aust. Vet. J.* **2011**, *89*, 6–10. [CrossRef]
48. Dynon, K.; Varrasso, A.; Ficorilli, N.; Holloway, S.; Reubel, G.; Li, F.; Hartley, C.; Studdert, M.; Drummer, H. Identification of equine herpesvirus 3 (equine coital exanthema virus), equine gammaherpesviruses 2 and 5, equine adenoviruses 1 and 2, equine arteritis virus and equine rhinitis A virus by polymerase chain reaction. *Aust. Vet. J.* **2001**, *79*, 695–702. [CrossRef] [PubMed]
49. Studdert, M.; Blackney, M. Equine herpesviruses: On the Differentiation of respiratory from foetal strains of type 1. *Aust. Vet. J.* **1979**, *55*, 488–492. [CrossRef]
50. Browning, G.F.; Studdert, M.J. Genomic Heterogeneity of Equine Betaherpesviruses. *J. Gen. Virol.* **1987**, *68*, 1441–1447. [CrossRef]
51. Turner, A.J.; Studdert, M.J.; Peterson, J.E. 2. Persistence of Equine Herpesviruses in Experimentally Infected Horses and the Experimental Induction of Abortion. *Aust. Vet. J.* **1970**, *46*, 90–98. [CrossRef] [PubMed]
52. Dynon, K. Pathogenesis of equine herpes virus type 2 (EHV2) infection and cellular processing of EHV2 glycoprotein. In *School of Veterinary Science 2010*; University of Melbourne: Melbourne, VIC, Australia, 2010.
53. Brault, S.A.; Bird, B.H.; Balasuriya, U.B.; MacLachlan, N.J. Genetic heterogeneity and variation in viral load during equid herpesvirus-2 infection of foals. *Vet. Microbiol.* **2011**, *147*, 253–261. [CrossRef]

54. Brault, S.A.; Blanchard, M.T.; Gardner, I.A.; Stott, J.L.; Pusterla, N.; Mapes, S.M.; Vernaua, W.; De Jongc, K.D.; MacLachlana, N.J The immune response of foals to natural infection with equid herpesvirus-2 and its association with febrile illness. *Vet. Immunol. Immunopathol.* **2010**, *137*, 136–141. [CrossRef]
55. Drummer, H.E.; Reubel, G.H.; Studdert, M.J. Equine gammaherpesvirus 2 (EHV2) is latent in B lymphocytes. *Arch. Virol.* **1996**, *141*, 495–504. [CrossRef] [PubMed]
56. Borchers, K.; Wolfinger, U.; Ludwig, H.; Thein, P.; Baxi, S.; Field, H.J.; Slater, J.D. Virological and molecular biological investigations into equine herpes virus type 2 (EHV-2) experimental infections. *Virus Res.* **1998**, *55*, 101–106. [CrossRef]
57. Reubel, G.H.; Crabb, B.S.; Studdert, M.J. Diagnosis of equine gammaherpesvirus 2 and 5 infections by polymerase chain reaction. *Arch. Virol.* **1995**, *140*, 1049–1060. [CrossRef]
58. Nordengrahn, A.; Rusvai, M.; Merza, M.; Ekstrom, J.; Morein, B.; Belak, S. Equine herpesvirus type 2 (EHV-2) as a predisposing factor for Rhodococcus equi pneumonia in foals: Prevention of the bifactorial disease with EHV-2 immunostimulating complexes. *Vet. Microbiol.* **1996**, *51*, 55–68. [CrossRef]
59. Fortier, G.: Richard, E.; Hue, E.; Fortier, C.; Pronost, S.; Pottier, D.; Lemaitre, L.; Lekeux, P.; Borchers, K.; Thiryg, E. Long-lasting airway inflammation associated with equid herpesvirus-2 in experimentally challenged horses. *Vet. J.* **2013**, *197*, 492–495. [CrossRef] [PubMed]
60. Murray, M.J.; Eichorn, E.S.; Dubovi, E.J.; Ley, W.B.; Cavey, D.M. Equine herpesvirus type 2: Prevalence and seroepidemiology in foals. *Equine Vet. J.* **1996**, *28*, 432–436. [CrossRef] [PubMed]
61. Balfou, H.H.J.; Holman, C.J.; Hokanson, K.M.; Lelonek, M.M.; Giesbrecht, J.E.; White, D.R.; Schmeling, D.O.; Webb, C.-H.; Cavert, W; Wang, D.H.; et al. A prospective clinical study of Epstein-Barr virus and host interactions during acute infectious mononucleosis. *J. Infect. Dis.* **2005**, *192*, 1505–1512. [CrossRef] [PubMed]
62. Brault, S.A.; Maclachlan, N.J. Equid gammaherpesviruses: Persistent bystanders or true pathogens? *Vet. J.* **2011**, *187*, 14–15. [CrossRef] [PubMed]
63. Foote, C.E.; Love, D.N.; Gilkerson, J.R.; Whalley, J.M. Detection of EHV-1 and EHV-4 DNA in unweaned Thoroughbred foals from vaccinated mares on a large stud farm. *Equine Vet. J.* **2004**, *36*, 341–345. [CrossRef] [PubMed]
64. Pusterla, N.; Leutenegger, C.M.; Wilson, W.D.; Watson, J.L.; Ferraro, G.L.; Madigan, J.E. Equine herpesvirus-4 kinetics in peripheral blood leukocytes and nasopharyngeal secretions in foals using quantitative real-time TaqMan PCR. *J. Vet. Diagn. Investig.* **2005**, *17*, 578–581. [CrossRef] [PubMed]
65. Pusterla, N.; Mapes, S.; Wilson, W.D. Use of viral loads in blood and nasopharyngeal secretions for the diagnosis of EHV-1 infection in field cases. *Veterinary Record.* **2008**, 728–729. [CrossRef] [PubMed]
66. Gilkerson, J.R.; Love, D.N.; Whalley, J.M. Epidemiology of equine herpesvirus abortion: Searching for clues to the future. *Aust. Vet. J.* **1998**. *76*, 675–676. [CrossRef]
67. Gilkerson, J.R.; Love, D.N.; Whalley, J.M. Incidence of equine herpesvirus 1 infection in thoroughbred weanlings on two stud farms. *Aust. Vet. J.* **2000**, *78*, 277–278. [CrossRef]
68. Edington, N.; Bridges, C.G.; Huckle, A. Experimental reactivation of equid herpesvirus 1 (EHV 1) following the administration of corticosteroids. *Equine Vet. J.* **1985**, *17*, 369–372. [CrossRef]
69. Slater, J.D.; Borchers, K.; Thackray, A.M.; Field, H.J. The trigeminal ganglion is a location for equine herpesvirus 1 latency and reactivation in the horse. *J. Gen. Virol.* **1994**, *75*, 2007–2016. [CrossRef]
70. Borchers, K.; Ebert, M.; Fetsch, A.; Hammond, T.; Sterner-Kock, A. Prevalence of equine herpesvirus type 2 (EHV-2) DNA in ocular swabs and its cell tropism in equine conjunctiva. *Vet. Microbiol.* **2006**, *118*, 260–266. [CrossRef] [PubMed]
71. Walton, A.H.; Muenzer, J.T.; Rasche, D.; Boomer, J.S.; Sato, B.; Brownstein, B.H.; Pachot, A.; Brooks, T.L.; Deych, E.; Shannon, W.D. Reactivation of multiple viruses in patients with sepsis. *PLoS ONE* **2014**, *9*, e98819.

Article

Investigation of Three Newly Identified Equine Parvoviruses in Blood and Nasal Fluid Samples of Clinically Healthy Horses and Horses with Acute Onset of Respiratory Disease

Nicola Pusterla [1,*], Kaitlyn James [2], Samantha Barnum [1] and Eric Delwart [3,4]

1 Department of Medicine and Epidemiology, School of Veterinary Medicine, University of California, Davis, CA 95616, USA; smmapes@ucdavis.edu
2 Department of Obstetrics, Gynecology and Reproductive Biology, Massachusetts General Hospital, Boston, MA 02114, USA; kaitlynej@gmail.com
3 Vitalant Research Institute, San Francisco, CA 94118, USA; Eric.Delwart@ucsf.edu
4 Department of Laboratory Medicine, University of California, San Francisco, CA 94118, USA
* Correspondence: npusterla@ucdavis.edu; Tel.: +1-530-752-1039

Simple Summary: The objective of the present study was to determine the molecular frequency of three recently identified parvoviruses (equine parvovirus hepatitis, equine parvovirus CSF and equine copivirus) in blood and respiratory secretions of 667 equids with acute onset of fever and respiratory signs and 87 clinically healthy horses. One hundred and seventeen sick horses tested qPCR-positive for at least one of the three parvoviruses. Ten clinically healthy horses tested qPCR-positive for one of the equine parvoviruses. The frequency of detection of the three equine parvoviruses was similar between sick and clinically healthy horses, suggesting that these newly characterized viruses do not appear to contribute to the clinical picture of equids with respiratory disease. In order to prove the clinical relevance of any of these newly identified equine parvoviruses, experimental challenge studies using pure, clonal inocula will be required.

Abstract: Three newly identified equine parvoviruses (equine parvovirus hepatitis (EqPV-H), equine parvovirus CSF (EqPV-CSF) and equine copivirus (Eqcopivirus)) have recently been discovered in horses with respiratory signs. However, the clinical impact of these three equine parvoviruses has yet to be determined. Nasal fluid samples and blood from 667 equids with acute onset of fever and respiratory signs submitted to a diagnostic laboratory were analyzed for the presence of common equine respiratory pathogens (equine influenza virus, equine herpesvirus-1/-4, equine rhinitis A and B virus, *S. equi* subspecies *equi*) as well as EqPV-H, EqPV-CSF and Eqcopivirus by qPCR. An additional 87 clinically healthy horses served as controls. One hundred and seventeen sick horses tested qPCR-positive for at least one of the three parvoviruses. Co-infections with common respiratory pathogens and parvoviruses were seen in 39 sick equids. All 87 clinically healthy horses tested qPCR-negative for all tested common respiratory pathogens and 10 healthy horses tested qPCR-positive for one of the equine parvoviruses. When the frequency of detection for EqPV-H, EqPV-CSF and Eqcopivirus of equids with respiratory signs was compared to that of clinically healthy horses, the difference was not statistically significant ($p > 0.05$), suggesting that the three recently identified equine parvoviruses do not contribute to the clinical picture of equids with respiratory disease.

Keywords: equine parvoviruses; equine parvovirus hepatitis; equine parvovirus CSF; equine copivirus; nasal fluid; blood; qPCR; sick equids; healthy horses

Citation: Pusterla, N.; James, K.; Barnum, S.; Delwart, E. Investigation of Three Newly Identified Equine Parvoviruses in Blood and Nasal Fluid Samples of Clinically Healthy Horses and Horses with Acute Onset of Respiratory Disease. *Animals* **2021**, *11*, 3006. https://doi.org/10.3390/ani11103006

Academic Editor: Margarita Martín Castillo

Received: 28 September 2021
Accepted: 14 October 2021
Published: 19 October 2021

Publisher's Note: MDPI stays neutral with regard to jurisdictional claims in published maps and institutional affiliations.

1. Introduction

Equine infectious respiratory diseases represent one of the most common clinical entities reported by practicing veterinarians nationwide [1], with equine influenza virus (EIV), equine herpesvirus-1 (EHV-1), EHV-4 and equine rhinitis A (ERAV) and B (ERBV)

viruses being considered the leading respiratory viruses [2–5]. The list of newly identified respiratory viruses in humans and various animal species has in the past decade expanded with the introduction of metagenomics [6,7]. This approach uses viral particle enrichment, random nucleic acid amplification and deep sequencing followed by bioinformatics analysis for the presence of viral sequences [8]. Two studies have recently reported three new equine parvoviruses, named equine parvovirus hepatitis (EqPV-H), equine parvovirus CSF (EqPV-CSF) and equine copivirus (Eqcopivirus) [9,10]. These equine parvoviruses were identified in blood and nasal secretions of apparently healthy horses and horses with acute onset of respiratory signs. However, these studies were unable to demonstrate causality for these newly identified equine parvoviruses. It is, therefore, the aim of this study to determine the frequency of genome detection of three newly identified parvoviruses (EqPV-H, EqPV-CSF and Eqcopivirus) in blood and nasal fluid samples of horses with acute onset of fever and respiratory signs, as well as clinically healthy control horses, and to determine potential demographic and clinical prevalence factors associated with these parvoviruses.

2. Materials and Methods

2.1. Study Population and Sampling

Blood and nasal fluid samples from 667 horses, mules and donkeys with acute onset of fever and respiratory signs were enrolled in the study. The samples were submitted to the Real-Time PCR Research and Diagnostics Core Facility, School of Veterinary Medicine, University of California at Davis from 1 January 2020 to 31 December 2020. Demographic and clinical information was gathered from the submission forms. Additional blood and nasal fluid samples from 87 clinically healthy horses were collected during the same period to determine the rate of equine parvoviruses in this population. The clinically healthy horses presented to the William R. Pritchard Veterinary Medical Teaching Hospital, School of Veterinary Medicine, University of California at Davis for routine health care (vaccination, health care certificate, oral examination) or elective procedures. Clients were asked to fill out a consent form allowing the collection of biological samples.

2.2. DNA Purification and Quantitative PCR Analyses

Nucleic acid extraction from whole blood and nasal fluid samples was performed using an automated nucleic acid extraction system (QIAcubeHT, Germantown, MD, USA) according to the manufacturer's recommendations. For blood samples, 100 µL of whole blood was processed for nucleic acid purification. The collected nasal swabs were placed into a conical tube containing 1000 µL of phosphate-buffered saline (PBS), vortexed and quickly centrifuged in order to release the nasal fluid from the swab. A total of 200 µL of nasal fluid/PBS solution was processed for nucleic acid purification.

Nasal fluid samples from sick and clinically healthy horses were tested for the presence of common respiratory pathogens, including EIV, EHV-1, EHV-4, equine rhinitis A and B virus (ERVs) and *S. equi*, as previously reported [2,11]. Blood from the same horses was tested for EHV-1 [2]. Further, nasal fluid samples and blood of all study horses were tested for EqPV-H, EqPV-CSF and Eqcopivirus using established and validated qPCR assays. Published sequences from GenBank (www.ncbi.nlm.nih.gov/genbank, accessed on 16 September 2019) of all three equine parvoviruses were subjected to BLAST analysis, and the aligned sequences were used to design primers and probes (Table 1). The samples were amplified in a combined thermocycler/fluorometer (7900 HT Fast, Applied Biosystems, Foster City, CA, USA) with the standard thermal cycling protocol: 2 min at 50 °C, 10 min at 95 °C and 40 cycles of 15 s at 95 °C and 60 s at 60 °C. The PCR reactions for each of the three parvovirus assays was composed of a commercially available mastermix (Universal TaqMan Mastermix with AmpErase UNG, Applied Biosystems, Foster City, CA, USA), containing 10 mM Tris (pH 8.3), 50 mM KCl, 5 mM $MgCl_2$, 300 µM each of dATP, dCTP and dGTP, 600 µM dUTP, 0.625 U of AmpliTaq Gold per reaction, 0.25 U AmpErase UNG per reaction, 400 nM of each primer and 80 nM of the respective TaqMan probe and 1 µL of DNA sample for a total volume of 12 µL. For each of the three parvoviruses, standard curves were generated using synthetic long oligonucleotides containing the target sequence for EqPV-H, EqPV-CSF and Eqcopivirus, and the amplification efficiency was calculated from the slope using the formula $E = 10^{[-1/slope]}$. The amplification efficiency was 99%, 100% and 95% for the capsid protein gene of EqPV-H, the capsid VP1 gene of EqPV-CSF and the NS1 gene of Eqcopivirus, respectively, indicating a very high analytical sensitivity. The detection limit for the three parvovirus assays was 13 genome equivalents when the DNA was purified from nasal fluid samples and whole blood. To determine the quality and efficiency of nucleic acid extraction, all samples were analyzed for the presence of the housekeeping gene equine glyceraldehyde-3-phosphate dehydrogenase (eGAPDH), as previously described [2].

Table 1. Oligonucleotide sequences of primers and probes used to detect three newly identified equine parvoviruses by qPCR.

Equine Parvovirus	Target Gene (GenBank)	Oligonucleotides
EqPV-H	Capsid protein (MH500792)	EqPV-H-forward primer: AGAATGCAGATGCTTTCCGAC EqPV-H-reverse primer: AAAGCAGATCCCGAATCCG EqPV-H-probe: FAM-GAAGATTCATGAGCTAGTC-MGB
EqPV-CSF	Capsid VP1 (KR902500)	EqPV-CSF-forward primer: AAGGCTTTGGACAAACGGG EqPV-CSF-reverse primer: TTGTTAGCACATGCGTTCCC EqPV-CSF-probe: FAM-AAGGGATATGGAAGGGA-MGB
Eqcopivirus	NS1 (MN181468)	EqCopi-forward primer: TCGCCCAGATCGTTGAGAAC EqCopi-reverse primer: AGCTGCTGTCTCCTGTTGTCC EqCopi-probe: FAM-ACCCAATCACCGAAGC-MGB

2.3. Statistical Analyses

Descriptive analyses (mean, standard deviation and median) were performed to evaluate the demographic and clinical information from the submission forms. Categorical analyses were performed using a Pearson's chi-square test to determine the association between observations (age, breed, sex, clinical signs (rectal temperature, nasal discharge, coughing) and infections. Each parvovirus infectious disease group was compared to non-parvovirus infected sick horses. To avoid interpretation bias when multiple pathogens were involved, only horses with a single pathogen (parvoviruses and non-parvoviruses) were evaluated. All statistical analyses were performed using commercial software (Stata Statistical Software, Version 14, College Station, TX, USA) and statistical significance was set at $p < 0.05$.

3. Results

Demographic and clinical information (age, breed and sex) was available for 453/667 sick equids (68%). The age of the sick equids ranged from 1 month to 34 years, with a median of 9 years. Sixty-one percent of the equids were males (stallions and geldings), and 39% were females. A variety of breeds were represented, including Quarter Horse (37%), Warmblood (14%), Thoroughbred (10%), pony breed (6%), Arabian (5%), Paint Horse (4%) and other breeds (22%). There were 12 donkeys and 3 mules (2%) reported. Clinical signs included fever (range 38.6 to 41.4 °C, median 39.4 °C) in 97%, nasal discharge in 74% and coughing in 46% of equids with reported clinical signs. The population of clinically healthy horses was composed of 50 males (57%) and 37 females (43%) with ages from 3 months to 32 years (median 7.5 years).

The frequency of detection for common respiratory pathogens in sick equids was as follows: 81 EIV (12.1%), 61 *S. equi* (9.1%), 50 EHV-4 (7.5%), 36 ERVs (5.4%) and 13 EHV-1 (1.9%, Table 2). Four equids EHV-1 qPCR-positive in nasal fluid samples also tested positive for EHV-1 in blood. Overall, 48 equids tested qPCR-positive for EqPV-H (2 nasal fluid samples only, 40 blood only, 6 both nasal fluid samples and blood). For EqPV-CSF, 35 equids tested qPCR-positive (4 nasal fluid samples only, 27 blood only, 4 both nasal fluid samples and blood). Fifty-nine equids tested qPCR-positive for Eqcopivirus (10 nasal fluid samples only, 24 blood only, 25 both nasal fluid samples and blood). Amongst the 117 equids with equine parvovirus infection, 95 had a single infection (46 Eqcopivirus, 29 EqPV-H and 20 EqPV-CSF), 20 had dual infections (10 EqPV-H and EqPV-CSF, 7 Eqcopivirus and EqPV-H and 3 Eqcopivirus and EqPV-CSF) and 2 had triple infections (Eqcopivirus, EqPV-H and EqPV-CSF). Co-infections with common respiratory pathogens and parvoviruses were seen in 39 equids (15 *S. equi*, 11 EIV, 11 ERVs, 8 EHV-4 and 1 EHV-1). All 87 clinically healthy horses tested qPCR negative for EIV, EHV-1, EHV-4, ERVs and *S. equi*. Ten clinically healthy horses tested qPCR positive for one of the equine parvoviruses (5 EqPV-H, 4 Eqcopivirus and 1 EqPV-CSF; Table 2).

Table 2. Frequency of detection of common respiratory pathogens and newly identified equine parvoviruses in sick and clinically healthy equids.

Pathogen	Sick Equids (667)		Clinically Healthy Horses (87)	
	Nasal Fluid	Blood	Nasal Fluid	Blood
EIV	81 (12.1%)	Not tested	0	Not tested
S. equi	61 (9.1%)	Not tested	0	Not tested
EHV-4	50 (7.5%)	Not tested	0	Not tested
ERVs	36 (5.4%)	Not tested	0	Not tested
EHV-1	13 (1.9%)	4 (0.6%)	0	0
EqPV-H	8 (1.2%)	46 (6.9%)	1 (1.1%)	5 (5.7%)
EqPV-CSF	8 (1.2%)	32 (4.8%)	0	1 (1.1%)
Eqcopivirus	35 (5.2%)	49 (7.3%)	2 (2.3%)	4 (4.6%)

When demographic prevalence factors were determined for each of the infectious groups (common respiratory pathogen and parvoviruses), the median age for EHV-4 and ERVs was significantly lower compared to most of the other groups ($p < 0.05$; Table 3). The EqPV-H and EqPV-CSF groups had the highest median age population with 10 and 15 years of age, respectively. There were more male than female equids in the various groups, although the differences were not statistically significant ($p > 0.05$). While the rectal temperature for the sick horses showed a wide range from 36.9 to 41.4 °C, there were no differences in median rectal temperatures amongst the various infectious groups. The reported frequency of nasal discharge for the various infectious groups ranged from 61.1% to 90.2%, with the lowest frequencies found in the three parvovirus groups. Equids infected with EIV had a significantly higher frequency of reported nasal discharge when compared to equids from the EHV-4, EqPV-H, EqPV-CSF, Eqcopivirus and the negative infection group ($p < 0.01$). The frequency of coughing ranged from 22.2% to 90.2%, with the highest frequency found in the EIV infection group and the lowest in the EHV-1 infection group. The EIV infection group had a significantly higher frequency of reported coughing compared to all the other infection groups, with the exception of the ERVs group ($p < 0.01$).

When the frequency of detection of EqPV-H, EqPV-CSF and Eqcopivirus, in plasma and nasal fluid samples of equids with respiratory signs was compared to that in plasma and nasal fluid samples of healthy horses, the difference was not statistically significant ($p > 0.05$).

Table 3. Demographic and clinical data from 667 equids with acute onset of fever and/or respiratory signs. The data is presented for each common respiratory pathogen and the three newly identified equine parvoviruses. Only animals with a single detected pathogen are reported. Demographic and clinical information was available for approximately 68% of the sick equids.

Pathogen	Demographic		Clinical Signs		
	Age Range in Years (Median)	Sex Distribution (Male/Female)	Rectal Temperature Range in °C (Median)	Nasal Discharge (%)	Coughing (%)
EHV-1 (9)	1–12 (5)	5/4	38.8–40.9 (39.8)	6/9 (66.7)	2/9 (22.2)
EHV-4 (35)	1–23 (4)	20/15	38.6–40.9 (39.6)	24/35 (68.6)	12/35 (34.3)
EIV (61)	1–22 (8)	28/33	38.6–41.1 (39.3)	55/61 (90.2)	55/61 (90.2)
ERVs (17)	1–25 (3)	9/8	38.6–40.8 (39.4)	13/17 (76.5)	10/17 (58.8)
S. equi (37)	2–22 (7)	23/14	38.7–41.0 (39.4)	31/37 (83.8)	16/37 (43.2)
Eqcopivirus (31)	1–30 (8.5)	16/15	38.1–41.4 (39.4)	20/31 (64.5)	14/31 (45.2)
EqPV-H (18)	1–30 (10)	11/7	38.0–40.6 (39.4)	11/18 (61.1)	8/18 (44.4)
EqPV-CSF (13)	1–26 (15)	8/5	36.9–40.0 (39.3)	8/13 (61.5)	7/13 (53.8)
Negative (376)	1–34 (9)	206/170	37.2–41.2 (39.4)	262/376 (69.7)	137/376 (36.4)

4. Discussion

Although the three newly identified equine parvoviruses have been characterized from biological samples of both clinically diseased and clinically healthy horses, their clinical relevance has remained elusive. The best-investigated equine parvovirus is EqPV-H, which has been linked to clinical and subclinical hepatitis [12–16]. EqPV-H has been shown to infect and replicate in hepatocytes, and viral infection is associated with liver pathology during hepatitis [11,15]. Further, EqPV-H has been detected in oral and nasal secretions and feces of experimentally infected horses, suggesting that these biological samples may be involved in horizontal transmission [15]. While EqPV-H has been experimentally transmitted via the oral route in one single horse, the intranasal inoculation of EqPV-H containing serum in two horses did not lead to viremia [15]. However, other transmission routes, such as the oral, nasal and vector-borne route, remain possible, especially around peak viremia [15,16]. Based on a recent study, vertical transmission does not appear to be a major contributor to the epidemiology of EqPV-H [14]. Reports on EqPV-CSF and Eqcopivirus have been sparse since the first description of these equine parvoviruses [9,10]. EqPV-CSF was initially found in the cerebrospinal fluid of a horse with neurological signs and has since then been reported in serum and nasal secretions of both clinically healthy horses and horses with fever and respiratory signs [9,10,17]. To the authors' knowledge, only one single report has documented the genomic presence of Eqcopivirus in blood and/or nasal secretions of healthy horses and horses with respiratory signs [10]. The detection of any of the three newly identified equine parvoviruses in nasal fluid samples is relevant, as viral shedding can contribute to environmental contamination with direct or indirect transmission to susceptible equids.

One hundred and seventeen horses (17%) with acute onset of fever and respiratory signs tested qPCR-positive for at least one of the three parvoviruses. Blood was the predominant sample positive for EqPV-H and EqPV-CSF, while blood and nasal fluid samples tested qPCR-positive for Eqcopivirus with similar frequencies. It appears that the three newly identified equine parvoviruses may have a species affinity, as all 15 donkeys and mules tested qPCR-negative for the three parvoviruses. While a large population of non-horse equids will need to be screened to strengthen the parvovirus species-specificity, the present study results are in agreement with a previous study, which was unable to detect EqPV-H in 13 donkeys from Austria [18].

One of the characteristics of a respiratory virus is its tropism to epithelial cells and the transient shedding in respiratory secretions during clinical or subclinical disease. Neither EqPV-H nor EqPV-CSF showed frequent detection by qPCR in nasal secretions, which agrees with a previous study [10]. The low detection rate of these two parvoviruses in nasal fluid samples reduces the likelihood that these two viruses are associated with clinical respiratory disease. Further, the frequency of detection of EqPV-H and EqPV-CSF in sick equids was similar to the one detected in clinically healthy horses. While there was no difference in the detection of Eqcopivirus between sick and clinically healthy equids, this virus had the highest detection rate in nasal fluid samples compared to the two other parvoviruses. Even if Eqcopivirus is not associated with clinical respiratory disease, the high detection rate observed in the present study population may relate to the spread of this virus via nasal secretions and possible transmission via droplets.

The study results showed that horses with active parvovirus infection were significantly older compared to horses infected with common respiratory pathogens. These results are in agreement with two recent studies, which determined that age was a risk factor influencing the rate of EqPV-H infections [18,19]. The study by Badenhorst and colleagues [18] determined that with every increase of 1 year in age, the risk of active EqPV-H infection was 1.1 times higher. Other demographic and clinical risk factors associated with any of the three equine parvoviruses were not any different than for horses presenting with fever, nasal discharge and coughing and having no common respiratory pathogens detected by qPCR. The similarity of prevalence factors between the parvovirus qPCR-positive equids and sick equids without detectable respiratory pathogens in nasal secretions further reinforces the lack of causality between these three equine parvoviruses and respiratory disease in the present study population.

The frequency of detection of the three parvoviruses in plasma and nasal fluid samples was comparable between horses with respiratory disease and clinically healthy horses. Further, the DNA prevalence for EqPV-H, EqPV-CSF and Eqcopivirus detected in the study population of sick and clinically healthy equids was similar to previously published studies. Surveillance studies in healthy horses from the USA, China, Austria and Germany have reported DNA prevalence for EqPV-H ranging between 7.1% and 17.0% [10,12,18–21]. The frequency of detection of EqPV-CSF in sick and clinically healthy horses was similar to the 4.9% detection rate reported in 41 healthy horses from the USA [10] and lower than the 23.1% reported from healthy Thoroughbred horses undergoing a custom quarantine in North Xinjiang province, China [17]. The prevalence of Eqcopivirus in plasma and or nasal fluid samples from clinically healthy equids in the present study was lower than the previously reported prevalence of 17% determined in plasma samples from apparently healthy horses [10]. Differences in horse populations used for the various studies are likely the reason for the observed differences in equine parvovirus detection rates.

Limitations of the study relate to the inability to re-sample the study animals. Longitudinal monitoring of affected equids would allow determining viral outcome following the convalescent period. One additional limitation was that the clinically healthy control population included in this study was not matched for age, time and location.

5. Conclusions

In conclusion, while EqPV-H, EqPV-CSF and Eqcopivirus can be found predominantly in the blood of equids with acute onset of fever and respiratory signs, it does not appear that these three newly identified parvoviruses contribute to the clinical picture of equids with respiratory disease. Since the initial characterization of EqPV-H, EqPV-CSF and Eqcopivirus, two additional equine parvoviruses and a previously unknown picornavirus have been described in the tissues of horses with interstitial pneumonia [22]. In order to prove the clinical relevance of any of these new equine parvoviruses, experimental challenge studies using pure, clonal inocula will be required.

Author Contributions: Conceptualization, N.P., K.J. and E.D.; data curation, N.P. and K.J.; formal analysis, S.B.; funding acquisition, N.P.; investigation, N.P. and S.B.; methodology, S.B.; project administration, N.P. and S.B.; resources, N.P.; software, N.P. and K.J.; supervision, N.P. and E.D.; validation, S.B.; visualization, N.P.; writing–original draft, N.P.; writing–review and editing, N.P., K.J., S.B. and E.D. All authors have read and agreed to the published version of the manuscript.

Funding: This research was funded by an Advancement in Equine Research Award, Boehringer Ingelheim Animal Health and by the Center for Equine Health, School of Veterinary Medicine, University of California, Davis, with additional contributions from public and private donors (grant number 20–17).

Institutional Review Board Statement: The study was approved by the Institutional Animal Care and Use Committee at the University of California at Davis (protocol code 21904, approved 9 February 2020).

Data Availability Statement: Data available on request due to privacy restrictions.

Acknowledgments: The authors would like to thank all equine veterinarians who submitted samples to the laboratory.

Conflicts of Interest: The authors declare no conflict of interest.

References

1. Traub-Dargatz, J.L.; Salman, M.D.; Voss, J.L. Medical problems of adult horses, as ranked by equine practitioners. *J. Am. Vet. Med. Assoc.* **1991**, *198*, 1745–1747. [PubMed]
2. Pusterla, N.; Kass, P.H.; Mapes, S.; Johnson, C.; Barnett, D.C.; Vaala, W.; Gutierrez, C.; McDaniel, R.; Whitehead, B.; Manning, J. Surveillance programme for important equine infectious respiratory pathogens in the USA. *Vet. Rec.* **2011**, *169*, 12. [CrossRef] [PubMed]
3. Burrows, R.; Goodridge, D.; Denyer, M.; Hutchings, G.; Frank, C.J. Equine influenza infections in Great Britain. 1979. *Vet. Rec.* **1982**, *110*, 494–497. [CrossRef] [PubMed]
4. Mumford, E.L.; Traub-Dargatz, J.L.; Salman, M.D.; Collins, J.K.; Getzy, D.; Carman, J. Monitoring and detection of acute viral respiratory tract disease in horses. *J. Am. Vet. Med. Assoc.* **1998**, *213*, 385–390.
5. Mumford, E.L.; Traub-Dargatz, J.L.; Carman, J.; Callan, R.J.; Collins, J.K.; Goltz, K.L.; Romm, S.R.; Tarr, S.F.; Salman, M.D. Occurrence of infectious upper respiratory tract disease and response to vaccination in horses on six sentinel premises in northern Colorado. *Equine Vet. J.* **2003**, *35*, 72–77. [CrossRef]
6. Chiu, C.Y. Viral pathogen discovery. *Curr. Opin. Microbiol.* **2013**, *16*, 468–478. [CrossRef]
7. Chiu, C.Y.; Miller, S.A. Clinical metagenomics. *Nat. Rev. Genet.* **2019**, *20*, 341–355. [CrossRef]
8. Breitwieser, F.P.; Lu, J.; Salzberg, S.L. A review of methods and databases for metagenomic classification and assembly. *Brief Bioinform.* **2019**, *20*, 1125–1136. [CrossRef]
9. Li, L.; Giannitti, F.; Low, J.; Keyes, C.; Ullmann, L.S.; Deng, X.; Aleman, M.; Pesavento, P.A.; Pusterla, N.; Delwart, E. Exploring the virome of diseased horses. *J. Gen. Virol.* **2015**, *96*, 2721–2723. [CrossRef]
10. Altan, E.; Li, Y.; Sabino-Santos, G., Jr.; Sawaswong, V.; Barnum, S.; Pusterla, N.; Deng, X.; Delwart, E. Viruses in horses with neurologic and respiratory diseases. *Viruses* **2019**, *11*, 942. [CrossRef]
11. Pusterla, N.; Mapes, S.; Wademan, C.; White, A.; Hodzic, E. Investigation of the role of lesser characterised respiratory viruses associated with upper respiratory tract infections in horses. *Vet. Rec.* **2013**, *172*, 315. [CrossRef]
12. Divers, T.J.; Tennant, B.C.; Kumar, A.; McDonough, S.; Cullen, J.; Bhuva, N.; Jain, K.; Chauhan, L.S.; Scheel, T.K.H.; Lipkin, W.I.; et al. New parvovirus associated with serum hepatitis in horses after inoculation of common biological product. *Emerg. Infect. Dis.* **2018**, *24*, 303–310. [CrossRef]
13. Tomlinson, J.E.; Tennant, B.C.; Struzyna, A.; Mrad, D.; Browne, N.; Whelchel, D.; Johnson, P.J.; Jamieson, C.; Löhr, C.V.; Bildfell, R.; et al. Viral testing of 10 cases of Theiler's disease and 37 in-contact horses in the absence of equine biologic product administration: A prospective study (2014–2018). *J. Vet. Intern. Med.* **2019**, *33*, 258–265. [CrossRef] [PubMed]
14. Tomlinson, J.E.; Kapoor, A.; Kumar, A.; Tennant, B.C.; Laverack, M.A.; Beard, L.; Delph, K.; Davis, E.; Schott Ii, H.; Lascola, K.; et al. Viral testing of 18 consecutive cases of equine serum hepatitis: A prospective study (2014–2018). *J. Vet. Intern. Med.* **2019**, *33*, 251–257. [CrossRef] [PubMed]
15. Tomlinson, J.E.; Jager, M.; Struzyna, A.; Laverack, M.; Fortier, L.A.; Dubovi, E.; Foil, L.D.; Burbelo, P.D.; Divers, T.J. Van de Walle, G.R. Tropism, pathology, and transmission of equine parvovirus-hepatitis. *Emerg. Microbes Infect.* **2020**, *9*, 651–663. [CrossRef] [PubMed]
16. Ramsauer, A.S.; Badenhorst, M.; Cavalleri, J.V. Equine parvovirus hepatitis. *Equine Vet. J.* **2021**, *53*, 886–894. [CrossRef] [PubMed]
17. Xie, J.; Tong, P.; Zhang, A.; Song, X.; Zhang, L.; Shaya, N.; Kuang, L. An emerging equine parvovirus circulates in thoroughbred horses in north Xinjiang, China, 2018. *Transbound. Emerg. Dis.* **2020**, *67*, 1052–1056. [CrossRef] [PubMed]
18. Badenhorst, M.; de Heus, P.; Auer, A.; Tegtmeyer, B.; Stang, A.; Dimmel, K.; Tichy, A.; Kubacki, J.; Bachofen, C.; Steinmann, E.; et al. Active equine parvovirus-hepatitis infection is most frequently detected in Austrian horses of advanced age. *Equine Vet. J.* **2021**. [CrossRef] [PubMed]
19. Meister, T.L.; Tegtmeyer, B.; Brüggemann, Y.; Sieme, H.; Feige, K.; Todt, D.; Stang, A.; Cavalleri, J.V.; Steinmann, E. Characterization of equine parvovirus in thoroughbred breeding horses from Germany. *Viruses* **2019**, *11*, 965. [CrossRef]
20. Lu, G.; Sun, L.; Ou, J.; Xu, H.; Wu, L.; Li, S. Identification and genetic characterization of a novel parvovirus associated with serum hepatitis in horses in China. *Emerg. Microbes Infect.* **2018**, *7*, 170. [CrossRef] [PubMed]

21. Lu, G.; Wu, L.; Ou, J.; Li, S. Equine parvovirus-hepatitis in China: Characterization of its genetic diversity and evidence for natural recombination events between the Chinese and American strains. *Front. Vet. Sci.* **2020**, *7*, 121. [CrossRef] [PubMed]
22. Altan, E.; Hui, A.; Li, Y.; Pesavento, P.; Asín, J.; Crossley, B.; Deng, X.; Uzal, F.A.; Delwart, E. New parvoviruses and picornavirus in tissues and feces of foals with interstitial pneumonia. *Viruses* **2021**, *13*, 1612. [CrossRef] [PubMed]

MDPI

Article

Seroprevalence of Equine Herpesvirus 1 (EHV-1) and Equine Herpesvirus 4 (EHV-4) in the Northern Moroccan Horse Populations

Zineb El Brini [1,*], Ouafaa Fassi Fihri [2], Romain Paillot [3], Chafiqa Lotfi [4], Farid Amraoui [4], Hanane El Ouadi [4], Mohamed Dehhaoui [5], Barbara Colitti [6], Hassan Alyakine [1] and Mohammed Piro [1]

1 Department of Medicine, Surgery, and Reproduction, Agronomy and Veterinary Institute Hassan II, Rabat 10000, Morocco; hassanalyakine@gmail.com (H.A.); vetpiro@yahoo.fr (M.P.)
2 Department of Microbiology, Immunology and Contagious Diseases, Agronomy and Veterinary Institute Hassan II, Rabat 10000, Morocco; fassifihri.ouafaa@gmail.com
3 School of Equine and Veterinary Physiotherapy, Writtle University College, Lordship Road, Writtle, Chelmsford CM1 3RR, UK; romain.paillot@writtle.ac.uk
4 Society of Veterinary Pharmaceutical and Biological Productions (Biopharma), Rabat 10000, Morocco; c.loutfi@biopharma.ma (C.L.); amraoui.farid@gmail.com (F.A.); hananelouadi@gmail.com (H.E.O.)
5 Department of Statistics and Applied Informatics, Agronomy and Veterinary Institute Hassan II, Rabat 10000, Morocco; dehhaoui@yahoo.fr
6 Department of Veterinary Sciences, University of Turin, Largo Braccini 2, 10095 Grugliasco, Italy; barbara.colitti@unito.it
* Correspondence: zinebelbrini@gmail.com; Tel.: +212-6-70-52-95-32

Citation: El Brini, Z.; Fassi Fihri, O.; Paillot, R.; Lotfi, C.; Amraoui, F.; El Ouadi, H.; Dehhaoui, M.; Colitti, B.; Alyakine, H.; Piro, M. Seroprevalence of Equine Herpesvirus 1 (EHV-1) and Equine Herpesvirus 4 (EHV-4) in the Northern Moroccan Horse Populations. *Animals* **2021**, *11*, 2851. https://doi.org/10.3390/ani11102851

Academic Editors: Amir Steinman and Oran Erster

Received: 5 July 2021
Accepted: 23 September 2021
Published: 29 September 2021

Publisher's Note: MDPI stays neutral with regard to jurisdictional claims in published maps and institutional affiliations.

Simple Summary: This work aims to evaluate the seroprevalence of equine EHV-1/4 in horse populations in the north of Morocco and to measure the antibody titers in vaccinated horses, under field conditions, with monovalent EHV-1 vaccines. Overall, 12.8% unvaccinated, and 21.8% vaccinated horses were positive for EHV-1. All samples were positive for EHV-4 when tested with the type-specific ELISA. The virus neutralization test showed low antibody titers in samples from vaccinated horses. Our study demonstrated that EHV-1 and EHV-4 are endemic in the horse populations in the north of Morocco and highlighted the necessity of reevaluating the vaccines and the vaccination protocol used.

Abstract: This study reports the first equine herpesvirus-1 (EHV-1) and equine herpesvirus-4 (EHV-4) seroprevalence investigation in horse populations of Morocco in 24 years. It also aims to determine antibody titers in horses vaccinated under field conditions with a monovalent EHV-1 vaccine. Blood samples were collected from 405 horses, including 163 unvaccinated and 242 vaccinated animals. They were tested using a commercial type-specific enzyme-linked immunosorbent assay (ELISA) and a virus neutralization test (VNT). Overall, 12.8% unvaccinated, and 21.8% vaccinated horses were positive for EHV-1. All samples were positive for EHV-4 when tested with the type-specific ELISA. In the vaccinated group, the VNT revealed a mean antibody titer of 1:49 for EHV-1 and 1:45 for EHV-4. The present study demonstrates that EHV-1 and EHV-4 are endemic in the horse populations in the north of Morocco, with prevalence differences between regions. Furthermore, horses vaccinated with a monovalent EHV-1 vaccine had low antibodies titers. This study highlights the necessity to establish and/or support efficient biosecurity strategies based on sound management of horses and characterize further and potentially improve the efficiency of the EHV vaccines and vaccination protocol used in the field.

Keywords: EHV-1; EHV-4; seroprevalence; ELISA; VNT; Morocco

1. Introduction

Equine herpesvirus 1 (EHV-1) and 4 (EHV-4) are common equine pathogens [1], causing significant economic losses and a negative impact on equine welfare [2]. EHV-1 and EHV-4 are closely related *Alphaherpesviruses* and, until 1981, were considered the same virus due to their genetic and antigenic similarity [3]. EHV-1 is associated with respiratory disease, abortion, neonate death, and equine herpesvirus myeloencephalopathy (EHM) [4], whereas EHV-4 is mainly related to respiratory disease, but can sporadically cause abortions [5]. The primary infection occurs through the upper respiratory tract by direct contact with respiratory secretions of actively infected horses, aborted fetuses, or placenta [6]. After the first infection, the virus establishes life-long latency (estimated to concern more than 80% of the cases), and reactivation can occur under natural conditions following transport, handling, postpartum period, or experimentally by treating horses with corticosteroids [2,7]. Consequentially, virus shedding could occur after reactivation from latency with a risk of spreading to susceptible animals.

In Morocco, the equine industry is essential for the country's socio-economic development, with a contribution of 0.61% to the country's GDP (Gross Domestic Product) and the direct and indirect employment of more than 30,000 people [8]. The Moroccan horse population is estimated at 110,000 horses, with around 4,000 births every year. Five main breeds are present; the Barb, the Arabian-Barb, the Arabian, Thoroughbred, and the Anglo-Arabian. The Arabian-Barb represents the majority, with 75 to 80% of the Moroccan horse population. To increase births and to reduce the losses of valuable horses, vaccination against EHV-1/4 has become a mandatory biosecurity practice required by the Moroccan authority since 2016. However, the obligation includes only breeding horses. At the same time, immunization is considered a practical approach when vaccinating a large population [9]. Moreover, vaccination efficiency in the field may vary depending on numerous factors, such as the level of virus strain circulation and/or the immune status at the time of vaccination and infection. Although EHV-1/4 vaccination reduces clinical signs of respiratory infection, virus shedding, and the occurrence of abortion storm, none of the available vaccines provide complete protection against all forms of the diseases, and none of them have been tested against EHM [5].

There is a paucity of information about the circulation of EHV-1 and EHV-4 in Morocco. The last available data come from a seroprevalence study conducted in 1997, using a virus neutralization test (VNT). This study reported an EHV-1/EHV-4 seroprevalence of 35% in tested horses [10]. Therefore, a better understanding of the EHV epidemiological situation is necessary, as it will play an essential role in preventing a disease that has a negative impact on horse welfare, breeding, and the equine sport industry. The recent EHV-1 outbreak in the CES Valencia (Spain) Spring Tours 2021 clearly illustrates the potentially devastating impact of EHV-1. Circulation of EHV-1 during this international show jumping competition that regrouped more than 750 horses has induced several hundred cases of infection, several deaths due to EHM, dissemination of the diseases in at least 9 European countries, and the subsequent cancellation of equestrian events in Europe by the FEI (Fédération Equestre Internationale) for several months (personal communication).

The development of a type-specific ELISA test, which is based on a type-specific epitope located at the C terminus of glycoprotein G (gG), represents an essential tool in the epidemiological investigation of EHV, allowing the specific sero-epizootiology and serodiagnosis of EHV-1 and EHV-4 [11]. The virus neutralization (VN) or the complement fixation (CF) tests are considered to be more cross-reactive, which tends to complicate results interpretation [12]. However, they are frequently used to assess the level of antibodies in response to a vaccination protocol. Heldens et al. [13] suggest that CF and VN antibodies may limit the duration of virus excretion, decrease the risk they pose to the other horses, and reduce the duration and severity of disease outbreaks.

The goals of this study were firstly to identify the seroprevalence of EHV-1 and EHV-4 using an EHV-1/4 type-specific ELISA in 405 sera from unvaccinated and vaccinated horses located in the provinces of Oujda, Meknes-Fez, Casablanca, El Jadida, and Marrakech. These provinces are located in the northern part of Morocco, which contains most of the horse population; and secondly to measure VN antibody titers in horses vaccinated with commercial monovalent inactivated EHV-1 vaccines currently used in Morocco, and to evaluate the serological status of the unvaccinated horses. This study is the first EHV-

1/4 serological investigation conducted in Morocco in more than two decades to better understand the EHV-1/4 epidemiological situation in the north of Morocco.

2. Materials and Methods

2.1. Area and Sampled Animals

This study was carried out on 405 horses, conveniently sampled and collected between March and May 2018, from 5 regions of Morocco that concentrate the largest population of horses in the north of Morocco and where national studs are located. The distribution of sera according to the sampling location is shown in Figure 1. 163 samples were taken from unvaccinated horses and 242 from horses vaccinated with commercially available inactivated monovalent EHV-1 vaccines. Horses were located in 5 provinces of northern Morocco (i.e., Oujda (n = 80; 32 unvaccinated), Meknes-Fez (n = 99; 34 unvaccinated), Casablanca (n = 83; all vaccinated), El Jadida (n = 62; 23 unvaccinated), and Marrakech (n = 81; 74 unvaccinated)).

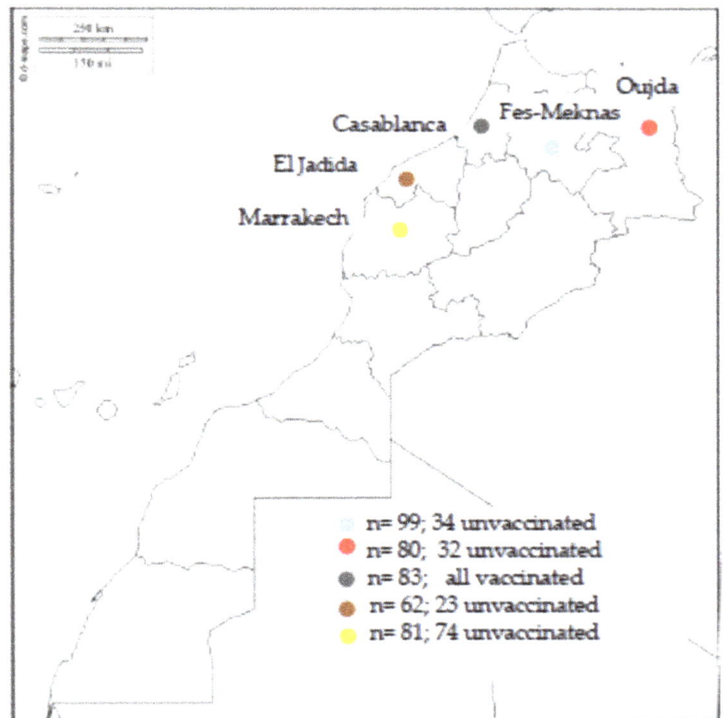

Figure 1. Geographical distribution of sampled regions in Morocco map.

The unvaccinated group was composed of 145 Arabian Barb, and 18 were from unknown breeds. There were 30% females (49/163) and 70% males (114/163) with ages ranging from 1 to 22 years (median age was 7.5 years). As the two main activities, 66% (107/163) of the unvaccinated individuals were working horses, while 34% (56/163) were horses involved in breeding.

For the vaccinated horses, 230 were Arabian Barb, and 12 were from unknown breeds. All of them were breeding horses as an activity, either stallion 20.6% (n = 50) or mares 80.2% (n = 192) with a median age of 8 years (range 4–20 years) at the time of sampling.

All immunized horses have been vaccinated by local veterinarians, under field conditions, according to the obligation of the Moroccan authorities (Royal Equestrian Society; SOREC) memorandum n° 95 of the 18/02/2015 following a preparatory course of two injections given 21 to 30 days apart. A third injection was given between 150 and 180 days (5 and 6 months) after the second injection, followed by annual boosters.

Vaccinated horses were enrolled in the study if they had received at least the primary vaccination and were vaccinated adequately with respect to the vaccination schedule. Two inactivated monovalent vaccines containing the EHV-1 Kentucky strain were used at the study time (Calvenza, Boehringer Ingelheim, Duluth, GA, USA or Pneumequine, Merial, Lyon, France) (Table 1).

Table 1. Information related to different vaccination parameters for the 242 vaccinated horses included in the study.

Vaccination Parameter	Category	Number of Individuals
Vaccination frequency	2 times	101
	3 times	81
	≥4 times	60
Days since last vaccination	1–90	85
	91–180	98
	≥181	59
Vaccine type *	A	170
	B	38
	C #	34

* A: Calvenza, B: Pneumequin, and C: Pneumequin and Calvenza # (# last vaccine used in 93% of the horses).

All the 405 sera samples were tested for EHV-1 and EHV-4 using the type-specific ELISA.

For comparison, the EHV-1 and EHV-4 VNT were performed on samples from unvaccinated and vaccinated horses based on the ELISA results obtained (see results section):

In the unvaccinated group, the EHV-1 VNT was performed for all EHV-1 ELISA positive sera and approximately half of the negative ones. The EHV-4 VNT was conducted on a total of 36 randomly selected individuals that had previously been tested as seronegative for EHV-1 and seropositive for EHV-4 by ELISA.

Regarding the vaccinated group, the EHV-1 VNT was performed for samples from 38 randomly selected individuals that had tested positive by EHV-1 ELISA, and another 64 samples from individuals randomly selected among the negative sera defined by EHV-1 ELISA. Additionally, within the group of vaccinated horses, the EHV-4 VNT was carried out with 50 samples randomly selected from EHV-1 ELISA seronegative horses.

All the horses were clinically healthy at the time of sampling with unknown history of infection with EHV-1 and/or EHV-4. The vaccines used were commercially available and regularly registered for equine species; no suffering was caused to the animals during the blood sampling. Owners were informed, and consent to the use of blood samples in this EHV seroprevalence study were obtained.

2.2. Samples Preparation

Blood samples were collected by jugular venipuncture into 10 mL vacutainer tubes without coagulant (Becton Dickinson, Le Pont De Claix, France). After clotting, the samples were centrifuged at 1500 rpm for 10 min. The sera aliquots were stored at −20 °C until further processing.

2.3. ELISA

According to the manufacturer's instruction, a commercial EHV-1 and EHV-4 diagnostic ELISA kit (Svanovir, Svanova AB, Uppsala, Sweden) were used to detect and discriminate EHV-1 and EHV-4 specific antibodies. The antibody values were detected by a 450 nm absorbance reading of each well. As indicated in the kit procedure, samples with optic density (OD) values > 0.20 were considered positive.

2.4. Virus Neutralization Test

Standard EHV VNT was performed as described in the Diagnostic Tests and Vaccines Manual for Terrestrial Animals [14]. Briefly, sera were inactivated at 56 °C for 30 min prior to the assay. The EHV-1 VNT was carried out in flat-bottomed 96 well sterile microtiter. Two-fold serial dilutions of sera were incubated with 100TCID50 per well of the EHV-1 Kentucky D strain MEM-5% FCS (Eagle's Medium were supplemented with 5% fetal calf serum). The plates were incubated at 37 °C in a 5% CO2 atmosphere for one hour before adding 10^5 rabbit kidney epithelial cells (RK-13, ATCC, CCL-37) in each well. The results were read microscopically after five days of culture. The highest dilution of serum resulting in 50% neutralization of virus was defined as the end-point titer. The test was validated with positive and negative serum controls and with back-titration of 100 $TCID_{50}$ doses. Neutralizing was calculated using the Karber Spearman formula. The EHV-4 VNT was performed exactly as described above using the EHV-4 405/75 strain and Equine dermis cells (ED, ATCC, CCL-57). Titers greater than or equal to 1:4 were considered positive [15].

Virus neutralization titers are presented after log10 transformation to allow a better comparison of results.

2.5. Statistical Analysis

The grouping of age was based on biological criteria (taking the consideration the life expectancy of the horses, i.e., 25–30 years approximately) and also based on the age distribution of the group of horses in the study itself in order to include to the extent the possible similar number of individuals within each category. A similar approach was used for grouping time since the last vaccination and the frequency of vaccination. We assessed the number of individuals so that each category includes the possible equal number of individuals to the extent.

All statistical assumptions were checked for normality and homogeneity of variances before performing any analyses. A chi-square (X^2) test was carried out to determine the association of the seroprevalence results with the different variables (vaccination status, regions, sex, age and activity, and the frequency of and the time of vaccination). An ANOVA test was performed to detect differences followed by a student t-test for mean comparison for multiple comparisons. Non-transformed titers were analyzed. p value < 0.05 was considered statistically significant. The IBM SPSS (version 25) and JMP (ver. 14.0.0) packages were used for statistical analysis.

3. Results

3.1. ELISA

A total of 405 sera samples were tested for EHV-1 and EHV-4 using the type-specific ELISA (Table S1).

All samples were found positive for the presence of the EHV-4 antibody. The EHV-1 seroprevalence was more variable. In the unvaccinated group, the EHV-1 seroprevalence was 12.9% (21/163) (95% confidence interval (CI), 8.16-19.02), whereas 21.1% (51/242) of sera collected from vaccinated horses were positive. The vaccinated and unvaccinated groups differed significantly for the EHV-1 seroprevalence response (X^2 = 4.470, p = 0.0345).

A statistically significant difference was found in the unvaccinated group considering the EHV-1 antibody prevalence between different regions (X^2 = 8.183 p = 0 042), El Jadida (30.4% 7/23), Fez-Meknes (5.9%, 2/34), Oujda (9.4%, 3/32), and Marrakech (12.1%, 9/74).

No significant effect was found regarding the sex (p = 0.730), the activity (either working or breeding horses) (p = 0.898), or the age (p = 0.256). However, in the unvaccinated horses, the incidence of the EHV-1 antibody increased significantly with age (p = 0.0172) (Table 2). In addition, there was no effect of the frequency (p = 0.718) or the time of the last vaccination ($p = 0.075$) on the seroprevalence of the EHV-1 antibody in the vaccinated horses.

Table 2. The difference for EHV-1 antibody values by the ELISA type-specific based on sex, activity, and age group of the horses included in the study.

		All Population		Non-Vaccinated		Vaccinated	
Variable	Df	X^2	p	X^2	p	X^2	p
Sex	1	0.552	0.4576	0.740	0.3897	1.300	0.2542
Activity	1	0.073	0.7876	2.070	0.1502	1.339	0.2472
Group of age	2	1.928	0.3814	8.125 *	0.0172 *	2.615	0.2705

* Significance for X^2 and p value. X^2: chi-square, DF: degrees of freedom for treatments, and p: probability.

3.2. Virus Neutralization Test

The EHV-1 and EHV-4 VNT were performed for samples from unvaccinated and vaccinated horses based on the ELISA results (Table S2) as described below:

Unvaccinated group: The results showed that 90.5% of EHV-1 ELISA positive sera were positive by VNT, with a mean antibody titer of 1:26 (4–95), while 53.6% (37/69) of the EHV-1 ELISA negative sera were positive by VNT with a mean antibody titer of 1:9 (4–24). The EHV-4 VNT revealed that 100% (36/36) of sera were positive. Their mean antibody titer was 1:19 (4–95).

Vaccinated group: The mean antibody titer was 1:49 (8–219) for EHV-1 positive and negative EHV-1 ELISA sera (no significant differences in their mean antibody titer ($p = 0.78$)) and 1:45 (8–166) for EHV-4. No significant difference in the mean antibody titer was found between EHV-1 and EHV-4 titers ($p = 0.51$). The ANOVA showed no effect of age and the frequency of vaccination on the values of the VNT. However, there was a significant effect on VNT values for the number of days since the last vaccination. VN values decreased when the number of days since the previous vaccination increased (Table 3). All sera from vaccinated horses were positive (titers > 1:4) for EHV-1 and EHV-4.

Table 3. Mean comparison of VNT dependent on the age, vaccination frequency, and the time since the last vaccination for the EHV-1 and EHV-4 combined.

Vaccination Parameter	Category	Number of Individuals	VN
Age (years)	1–6	63	44
	7–10	72	49
	≥11	24	52
Vaccination frequency	2 times	78	51
	3 times	59	45
	≥4 times	22	47
Days since last vaccination	1–90	73	56a
	91–180	59	42b
	≥181	27	36b

Means accompanied by different letters under the same column differ significantly for $\alpha = 0.01$.

4. Discussion

This study represents the first EHV-1 and the EHV-4 seroprevalence investigation conducted in the Moroccan horse populations in the last 24 years. Samples were collected from both vaccinated and unvaccinated, working and breeding horses located in five different regions of the north of Morocco. Serum was analyzed with the type-specific ELISA and the EHV-1/4 VNT.

4.1. Type-Specific ELISA

This study showed an overall EHV-1 seroprevalence rate of 12.8% in unvaccinated horses, while 100% of samples were positive for EHV-4. The high EHV-4 seroprevalence could be explained by an endemic circulation of EHV-4 with recurrent infection during the horse lifetime, inducing the antibody response to reach a plateau level [16]. EHV-4 outbreaks can occur all year round, with no link to seasonal variations, whereas EHV-1 outbreaks are usually reported in winter or early spring [17,18]. Moreover, Crabb et al. [19] suggest that the reactivation and/or reinfection with EHV-1 is less common. Consequently, the antibody response probably declines over time.

Interestingly, the EHV-1 seroprevalence in the vaccinated group was only 21.1% (51/242), regardless of time since or the frequency of vaccination. Despite that, all horses have received at least a primary course of vaccination. Our results suggest that the commercial type-specific ELISA could not reliably detect the antibody response produced by the EHV-1 vaccines used in Morocco. A study conducted by Yasunaga et al. [20,21] reported no difference in the antibody titer using a gG ELISA compared with the CF that revealed a significant increase in antibody titer after repeated intramuscular or intranasal vaccinations with an inactivated EHV-1 vaccine. In contrast, the study from Crabb et al. [19] reported that the type-specific ELISA was sensitive enough to detect a gG-specific antibody response after vaccination with an inactivated EHV-1/4 vaccine. The sensitivity of the gG ELISA might explain the difference. Crabb et al. [19] used a serum dilution of 1/1000, while the one used in the current study required a dilution of 1/10,000. The 100% seropositivity for EHV-4, in the vaccinated horses is likely to represent the seroprevalence of EHV-4 infection in the northern Moroccan horse population. However, in the absence of an EHV vaccine with DIVA capacity (Differentiating Infected from Vaccinated Animals), it is difficult to conclude if the current study's seropositive results obtained with the gG ELISA are linked to vaccination and/or natural infection.

There was a significant difference between the EHV-1 seroprevalence in the unvaccinated group (12.9%) when compared with the vaccinated populations (21.1%) by the ELISA test. Statistical analyses showed no effect of the frequency or the time since the last vaccination on the seroprevalence of the EHV-1. While not statistically significant in the current study, the horse sex may need to be considered. The majority of the vaccinated horses were mares (80.2%, 194/242), with breeding as the main activity (90.5%, 219/242), while the unvaccinated ones were primarily working horses (62%, 101/163) and male (69.9%, 114/163). It has been demonstrated that breeding mares are the principal reservoir of EHV-1 [22]. They undergo significant stress around the breeding and weaning period, resulting in more frequent reactivation of latent infections [23,24].

Numerous epidemiological investigations have been performed to measure EHV-1/4 seroprevalence worldwide. In our study, the overall EHV-1 and EHV-4 seroprevalences were 12.9% and 100%, respectively. In Morocco, previous studies have reported a seroprevalence between 32.38% and 51.5% for EHV-1 using VN and CF tests [10,25–27]. The strong cross-reaction might explain the difference in seroprevalence between EHV-1 and EHV-4 [12]. While similar seroprevalence using the same type-specific ELISA for EHV-1 18.8, 30, 23.2, and 21.1% and EHV-4 98.7, 100, 78, and 100% were reported, respectively by Dunowska et al. (New Zealand) [28], Ataseven et al. (Turkey) [29], Sáen et al. (Colombia) [30], and Crabb et al. (Australia) [19]. In a study conducted in Israel, a similar seroprevalence (99%) to EHV-4 was reported, with a very low seroprevalence (1%) to EHV-1 [31].

The results of the current study show an essential variation between regions. The higher EHV-1 seroprevalence was observed in El Jadida. This region encompasses the largest number of equids; breeding activity/farms, commingling, competition (racing, fantasia), and transportation of horses. These factors were identified as significant risks for the circulation of EHV-1 in horses [24]. When compared with another study [32], no climatic effect was associated with the regional seroprevalence of EHV-1 as all studied regions have a Mediterranean climate with only slight seasonal variations. Moreover, the lowest seroprevalence was in regions characterized by a cold winter, which has been identified as a stressor factor for EHV reactivation [17,33]. According to some studies, an increased incidence of EHV-1 seropositivity was observed in relation to the age in the unvaccinated horses [32,34]. Paillot et al. [35] reported that cell-mediated immunity to EHV-1 increased with age, which could be linked to repeated reactivation of latent EHV-1, infection, and vaccination. This result was not observed in the vaccinated horses. The effect of the vaccination might explain this difference, with the EHV-1 vaccine administered at an early age, potential frequent vaccination, and impact on infection/re-infection. However, this hypothesis needs to be confirmed in a more significant number of horses.

In contrast to the current study, the study conducted in Morocco by Hmidouch et al. in 1997 [10] found no effect of the horse density on the geographical distribution of the EHV-1/4 prevalence. The highest prevalence was observed in the region of Marrakech (39.07%), while the lowest prevalence was reported in the regions of El Jadida–Casablanca (24.93%). On the other hand, similar to our results, the sex and the age of the animal had no impact on the seroprevalence of the EHV1/4.

4.2. Virus Neutralization and ELISA Test

The neutralization antibody titers measured against EHV-1/4 in unvaccinated horses support a previous exposure and an active circulation of these viruses in horse populations of the north of Morocco. This result highlights the importance of this group as a source of infection and contamination for naive horses. We also found that the samples found negative when tested with the EHV-1 ELISA were, in fact, positive when tested with the VNT. Considering the high sensitivity and specificity of the type-specific ELISA as reported in previous studies [12,15,19], this result may be explained by the cross-reactivity between the EHV-1 and EHV-4 due to their antigenic similarity. In contrast, the EHV-4 VNT shows that all sera were positive in accordance with EHV-4 ELISA. These results strongly prove that EHV-4 is a ubiquitous virus actively circulating in horse populations in the north of Morocco.

In vaccinated horses, the aim of the VN assay was mainly to evaluate the antibody titers induced by an EHV-1 monovalent vaccine administered in field conditions. Our mean antibody titer was 1:49 for the EHV-1 and 1:45 for the EHV-4. Direct comparisons with other studies cannot be easily made, as our means were calculated on horses that were vaccinated on different days of the schedule of the vaccination program. However, relying only on the time since the last vaccination, even in the group that was vaccinated less than 90 days before sampling, the antibody titers remain low in comparison to other studies using an inactivated EHV vaccine (1:137-2048) [13,36,37] or the modified live (1:115-2048) [36,37]. This difference can be related to different factors. This is mainly due to the difference in the type of the vaccines, vaccination schedules, and the vaccine status at the time of vaccination. Indeed, Bannai et al. [38] suggest an effect of the previous infection with the EHV-4, which is antigenically cross-reactive with EHV-1 and could limit the increase in the antibody titer following vaccination. Attili et al. [39] suggested that the vaccine administration in animals with high antibody titers due to infection or previous vaccination could induce a decrease in antibody titer due to an interaction between antibodies and the vaccine. Based on the ELISA results, all our horses were positive to EHV-4.

There was no measurable effect of age or the frequency of the vaccination on the levels of the antibody titers. However, there was an inversely proportional relationship between the time of the vaccination and the VN antibodies titer; the fewer days between the time of vaccination and the sampling, the higher the VN values. This result was also reported in other studies, where the antibody titer started to decline 3 to 6 months after the vaccination [40–42]. Consequently, the approved vaccination protocol in Morocco may need to be reevaluated in order to incorporate more regular boost immunization for better protection.

In Morocco, horses were vaccinated with a monovalent EHV-1 vaccine to gain immunity for both viruses based on their genetic similarity. However, Lang et al. [15] revealed that even in natural infection, the increase in antibodies to the other virus was insufficient to generate a considerable seroconversion as the complete DNA sequence has proven significant genetic differences between the two Alphaherpesvirus [43,44]. The results of our study revealed no statistically significant difference in the average antibody titer against both viruses. This result may partly be explained by detecting cross-reactive antibodies when using the VNT, as previously demonstrated by Hartley et al. [12]. Moreover, Heldens et al. [13] suggested that monovalent vaccines would not offer adequate protection against heterologous challenges, and it would be unwise to rely on cross-protection. As a consequence, the use of the monovalent EHV-1 vaccine in Morocco may need to be reevaluated.

Finally, it is worth mentioning that the Arabian barb breed was overrepresented in the population sampled (i.e., 90.4%; due to the sampling process, availability, and selection) when compared with the overall Moroccan horse breed distribution (75 to 80%). Another potential limitation of the study is the imbalanced number between the group of vaccinated and unvaccinated in each region and the limited number of horses tested by the VNT. Therefore, our study only provides a snapshot of the situation and may not entirely represent Morocco's horse population.

5. Conclusions

EHV-1 and EHV-4 are endemic in horse populations in the north of Morocco. The EHV-1/4 type-specific ELISA revealed that all the horses were seropositive to EHV-4, while the seroprevalence of EHV-1 was more related to the region of origin. On the other hand, our results demonstrated that horses vaccinated in field conditions with a monovalent inactivated EHV-1 have a low level of antibody titers. An inversely proportional relationship was observed between the time since the last vaccination and the VN antibody titer. Considering these results, and the low frequency of vaccinated horses with measurable antibody titers, the vaccine and/or the vaccination schedule may need to be reevaluated. Epidemiology studies looking at the prevalence of EHV-1 and EHV-4 infection in Morocco will be necessary to confirm the level of EHV circulation and protection induced by vaccination. Moreover, further investigations will also be required to determine the annual losses due to EHV-1/4 in Morocco.

Supplementary Materials: The following are available online at https://www.mdpi.com/article/10.3390/ani11102851/s1, Table S1: Data—ELISA EHV Morocco, Table S2: Data—VNT EHV Morocco.

Author Contributions: Conceptualization, Z.E.B. and M.P.; methodology, C.L., H.E.O. and B.C.; software, M.D.; validation, M.P., R.P. and O.F.F.; formal analysis, Z.E.B.; investigation, Z.E.B.; data curation, Z.E.B.; writing—original draft preparation, Z.E.B.; writing—review and editing, Z.E.B., R.P. and M.P.; visualization, H.A.; supervision, M.P.; project administration, F.A. and M.P. All authors have read and agreed to the published version of the manuscript.

Funding: This work was supported by the Royal Equestrian Society (Société Royale d'Encouragement du Cheval, SOREC) and Agronomy and Veterinary Institute Hassan II.

Institutional Review Board Statement: The Royal Equestrian Society approved the study protocol (Société Royale d'Encouragement du Cheval, SOREC).

Informed Consent Statement: Owners gave consent for their animals' inclusion.

Data Availability Statement: The data presented in this study are available in Tables S1 and S2.

Acknowledgments: The authors are grateful to Sergio Rosati for the donation of the cells and the viruses and to the Society of Veterinary Biological and Pharmaceutical Products (Biopharma) for using their facilities, equipment, and assistance.

Conflicts of Interest: All authors declare no conflicts of interest.

References

1. Ma, G.; Azab, W.; Osterrieder, N. Equine Herpesviruses Type 1 (EHV-1) and 4 (EHV-4)—Masters of Co-Evolution and a Constant Threat to Equids and Beyond. *Vet. Microbiol.* **2013**, *167*, 123–134. [CrossRef]
2. Allen, G.P.; Kydd, J.H.; Slater, J.D.; Smith, K.C. Equid herpesvirus 1 (EHV-1) and equid herpesvirus 4 (EHV-4) infections. In *Infectious Diseases of Livestock*; Coetzer, J.A.W., Tustin, R.C., Eds.; Oxford Press: Cape Town, South Africa, 2004; pp. 829–859, Chapter 76.
3. Studdert, M.; Simpson, T.; Roizman, B. Differentiation of Respiratory and Abortigenic Isolates of Equine Herpesvirus 1 by Restriction Endonucleases. *Science* **1981**, *214*, 562–564. [CrossRef]
4. Laval, K.; Poelaert, K.C.K.; van Cleemput, J.; Zhao, J.; Vandekerckhove, A.P.; Gryspeerdt, A.C.; Garré, B.; van der Meulen, K.; Baghi, H.B.; Dubale, H.N.; et al. The Pathogenesis and Immune Evasive Mechanisms of Equine Herpesvirus Type 1. *Front. Microbiol.* **2021**, *12*, 662686. [CrossRef]
5. Slater, J. Equine herpesviruses. In *Equine Infectious Diseases*; Sellon, D.C., Long, M.T., Eds.; Saunders Elsevier: St. Louis, MO, USA, 2007; pp. 134–153.
6. Patel, J.R.; Heldens, J. Equine Herpesviruses 1 (EHV-1) and 4 (EHV-4)—Epidemiology, Disease and Immunoprophylaxis: A Brief Review. *Vet. J.* **2005**, *170*, 14–23. [CrossRef] [PubMed]
7. Borchers, K.; Wolfinger, U.; Ludwig, H. Latency-Associated Transcripts of Equine Herpesvirus Type 4 in Trigeminal Ganglia of Naturally Infected Horses. *J. Gen. Virol.* **1999**, *80*, 2165–2171. [CrossRef] [PubMed]
8. Filière équine. L'écosystème se développe à grande vitesse. LesEco.ma. 2019. Available online: https://leseco.ma/business/filiere-equine-l-ecosysteme-se-developpe-a-grande-vitesse.html (accessed on 30 June 2021). (In French).
9. Khusro, A.; Aarti, C.; Rivas-Caceres, R.R.; Barbabosa-Pliego, A. Equine Herpesvirus-I Infection in Horses: Recent Updates on Its Pathogenicity, Vaccination, and Preventive Management Strategies. *J. Equine Vet. Sci.* **2020**, *87*, 102923. [CrossRef] [PubMed]
10. Hmidouch, A.; Harrak, M.E.; Chakri, A.; Ouragh, L.; Lotfi, C.; Bakkali-Kassimi, L. Epidemiological study of equines rhinopneumonia in Morocco. *Rev. Élev. Méd. Vét. Pays Trop.* **1997**, *50*, 191–196. [CrossRef]
11. Crabb, B.S.; Studdert, M.J. Epitopes of Glycoprotein G of Equine Herpesviruses 4 and 1 Located near the C Termini Elicit Type-Specific Antibody Responses in the Natural Host. *J. Virol.* **1993**, *67*, 6332–6338. [CrossRef]
12. Hartley, C.A.; Wilks, C.R.; Studdert, M.J.; Gilkerson, J.R. Comparison of Antibody Detection Assays for the Diagnosis of Equine Herpesvirus 1 and 4 Infections in Horses. *Am. J. Vet. Res.* **2005**, *66*, 921–928. [CrossRef]
13. Heldens, J.G.M.; Hannant, D.; Cullinane, A.A.; Prendergast, M.J.; Mumford, J.A.; Nelly, M.; Kydd, J.H.; Weststrate, M.W.; van den Hoven, R. Clinical and Virological Evaluation of the Efficacy of an Inactivated EHV1 and EHV4 Whole Virus Vaccine (Duvaxyn EHV1,4). Vaccination/Challenge Experiments in Foals and Pregnant Mares. *Vaccine* **2001**, *19*, 4307–4317. [CrossRef]
14. World Organization for Animal Health (OIE). Chapter 2.5.9, Equine rhinopneumonitis (Infection with equid herpesvirus-1 And -4). In: OIE Terrestrial Manual 2017. Available online: https://www.oie.int/fileadmin/Home/eng/Health_standards/tahm/2.05.09EQUINE_RHINO.pdf. (accessed on 13 October 2020).
15. Lang, A.; de Vries, M.; Feineis, S.; Müller, E.; Osterrieder, N.; Damiani, A.M. Development of a Peptide ELISA for Discrimination between Serological Responses to Equine Herpesvirus Type 1 and 4. *J. Virol. Methods* **2013**, *193*, 667–673. [CrossRef] [PubMed]
16. Van Maanen, C. Equine Herpesvirus 1 and 4 Infections: An Update. *Vet. Q.* **2002**, *24*, 57–78. [CrossRef]
17. Matsumura, T.; Sugiura, T.; Imagawa, H.; Fukunaga, Y.; Kamada, M. Epizootiological Aspects of Type 1 and Type 4 Equine Herpesvirus Infections among Horse Populations. *J. Vet. Med. Sci.* **1992**, *54*, 207–211. [CrossRef] [PubMed]
18. DEUXIÈME APPEL À LA VIGILANCE: FOYERS D'HERPÈSVIROSES TYPE 1 (HVE1) AU 12 AVRIL. 2018. Available online: https://respe.net/deuxieme-appel-a-la-vigilance-foyers-dherpesviroses-type-1-hve1-au-12-avril-2018/ (accessed on 30 June 2021). (In French).
19. Crabb, B.S.; MacPherson, C.M.; Reubel, G.H.; Browning, G.F.; Studdert, M.J.; Drummer, H.E. A Type-Specific Serological Test to Distinguish Antibodies to Equine Herpesviruses 4 and 1. *Arch. Virol.* **1995**, *140*, 245–258. [CrossRef] [PubMed]
20. Yasunaga, S.; Maeda, K.; Matsumura, T.; Kai, K.; Iwata, H.; Inoue, T. Diagnosis and Sero-Epizootiology of Equine Herpesvirus Type 1 and Type 4 Infections in Japan Using a Type-Specific ELISA. *J. Vet. Med. Sci.* **1998**, *60*, 1133–1137. [CrossRef] [PubMed]
21. Yasunaga, S.; Maeda, K.; Matsumura, T.; Kondo, T.; Kai, K. Application of a Type-Specific Enzyme-Linked Immunosorbent Assay for Equine Herpesvirus Types 1 and 4 (EHV-1 and -4) to Horse Populations Inoculated with Inactivated EHV-1 Vaccine. *J. Vet. Med. Sci.* **2000**, *62*, 687–691. [CrossRef]
22. Gilkerson, J.R.; Whalley, J.M.; Drummer, H.E.; Studdert, M.J.; Love, D.N. Epidemiological Studies of Equine Herpesvirus 1 (EHV-1) in Thoroughbred Foals: A Review of Studies Conducted in the Hunter Valley of New South Wales between 1995 and 1997. *Vet. Microbiol.* **1999**, *68*, 15–25. [CrossRef]
23. Allen, G.P.; Bryans, J.T. Molecular Epizootiology, Pathogenesis, and Prophylaxis of Equine Herpesvirus-1 Infections. *Prog. Vet. Microbiol. Immunol.* **1986**, *2*, 78–144.
24. Lunn, D.P.; Davis-Poynter, N.; Flaminio, M.J.B.F.; Horohov, D.W.; Osterrieder, K.; Pusterla, N.; Townsend, H.G.G. Equine Herpesvirus-1 Consensus Statement. *J. Vet. Int. Med.* **2009**, *23*, 450–461. [CrossRef]
25. Ourahma, E. Contribution to the Epidemiological Study of Viral Respiratory Diseases of Equines in Morocco. Ph.D. Thesis, Hassan II Institute of Agronomy and Veterinary Medicine (IAV), Rabat, Morocco, 1984.
26. Lahlou-Kassi, S. Epidemiological Investigation on Infectious Anemia, Rhinopneumonia and Equine Virus Arteritis in Morocco. Ph.D. Thesis, Hassan II Institute of Agronomy and Veterinary Medicine (IAV), Rabat, Morocco, 1977.

27. Himer, D. Incidence in Morocco of Infectious Anemia, Rhinopneumonia and Viral Arteritis in Equines: Epidemiological Investigation—Prophylaxis. Ph.D. Thesis, National Veterinary School, Maisons-Alfort, France, 1975.
28. Dunowska, M.; Gopakumar, G.; Perrott, M.R.; Kendall, A.T.; Waropastrakul, S.; Hartley, C.A.; Carslake, H.B. Virological and Serological Investigation of Equid Herpesvirus 1 Infection in New Zealand. *Vet. Microbiol.* **2015**, *176*, 219–228. [CrossRef]
29. Ataseven, V.S.; Dağalp, S.B.; Güzel, M.; Başaran, Z.; Tan, M.T.; Geraghty, B. Prevalence of Equine Herpesvirus-1 and Equine Herpesvirus-4 Infections in Equidae Species in Turkey as Determined by ELISA and Multiplex Nested PCR. *Res. Vet. Sci.* **2009**, *86*, 339–344. [CrossRef] [PubMed]
30. Sáenz, J.R.; Góez, Y.; Inchima, S.U.; Góngora, A. Serologic evidence of equine herpesvirus 1 and 4 infection in two regions of Colombia. *Rev. Colomb. Cienc. Pecu.* **2008**, *21*, 251–258.
31. Aharonson-Raz, K.; Davidson, I.; Porat, Y.; Altory, A.; Klement, E.; Steinman, A. Seroprevalence and Rate of Infection of Equine Influenza Virus (H3N8 and H7N7) and Equine Herpesvirus (1 and 4) in the Horse Population in Israel. *J. Equine Vet. Sci.* **2014**, *34*, 828–832. [CrossRef]
32. Cruz, F.; Fores, P.; Mughini-Gras, L.; Ireland, J.; Moreno, M.A.; Newton, J.R. Seroprevalence and Factors Associated with Equine Herpesvirus Type 1 and 4 in Spanish Purebred Horses in Spain. *Vet. Rec.* **2016**, *178*, 398. [CrossRef]
33. Welch, H.M.; Bridges, C.G.; Lyon, A.M.; Griffiths, L.; Edington, N. Latent Equid Herpesviruses 1 and 4: Detection and Distinction Using the Polymerase Chain Reaction and Co-Cultivation from Lymphoid Tissues. *J. Gen. Virol.* **1992**, *73*, 261–268. [CrossRef]
34. Liutkevičien, V.; Stankevicien, M.; Mockeliunien, V.; Mockeliunas, R. Equine Herpes Viruses' Prevalence in Horse Population in Lithuania. *Biotechnol. Biotechnol. Equip.* **2006**, *20*, 111–115. [CrossRef]
35. Paillot, R.; Case, R.; Ross, J.; Newton, R.; Nugent, J. Equine Herpes Virus-1: Virus, Immunity and Vaccines. *Open Vet. Sci. J.* **2008**, *2*, 68–91. [CrossRef]
36. Goehring, L.S.; Wagner, B.; Bigbie, R.; Hussey, S.B.; Rao, S.; Morley, P.S.; Lunn, D.P. Control of EHV-1 Viremia and Nasal Shedding by Commercial Vaccines. *Vaccine* **2010**, *28*, 5203–5211. [CrossRef]
37. Goodman, L.; Wagner, B.; Flaminio, M.; Sussman, K.; Metzger, S.; Holland, R.; Osterrieder, N. Comparison of the Efficacy of Inactivated Combination and Modified-Live Virus Vaccines against Challenge Infection with Neuropathogenic Equine Herpesvirus Type 1 (EHV-1). *Vaccine* **2006**, *24*, 3636–3645. [CrossRef]
38. Bannai, H.; Tsujimura, K.; Nemoto, M.; Ohta, M.; Yamanaka, T.; Kokado, H.; Matsumura, T. Epizootiological Investigation of Equine Herpesvirus Type 1 Infection among Japanese Racehorses before and after the Replacement of an Inactivated Vaccine with a Modified Live Vaccine. *BMC Vet. Res.* **2019**, *15*, 280. [CrossRef]
39. Attili, A.-R.; Colognato, R.; Preziuso, S.; Moriconi, M.; Valentini, S.; Petrini, S.; de Mia, G.M.; Cuteri, V. Evaluation of Three Different Vaccination Protocols against EHV1/EHV4 Infection in Mares: Double Blind, Randomized Clinical Trial. *Vaccines* **2020**, *8*, 268. [CrossRef]
40. Wagner, B.; Goodman, L.B.; Babasyan, S.; Freer, H.; Torsteinsdóttir, S.; Svansson, V.; Björnsdóttir, S.; Perkins, G.A. Antibody and Cellular Immune Responses of Naïve Mares to Repeated Vaccination with an Inactivated Equine Herpesvirus Vaccine. *Vaccine* **2015**, *33*, 5588–5597. [CrossRef] [PubMed]
41. Heldens, J.G.M.; Kersten, A.J.; Weststrate, M.W.; van den Hoven, R. Vaccinology: Duration of Immunity Induced by an Adjuvanted and Inactivated Equine Influenza, Tetanus and Equine Herpesvirus 1 and 4 Combination Vaccine. *Vet. Q.* **2001**, *23*, 210–217. [CrossRef]
42. Hannant, D.; Jessett, D.M.; O'Neill, T.; Dolby, C.A.; Cook, R.F.; Mumford, J.A. Responses of Ponies to Equid Herpesvirus-1 Iscom Vaccination and Challenge with Virus of the Homologous Strain. *Res. Vet. Sci.* **1993**, *54*, 299–305. [CrossRef]
43. Telford, E.A.R.; Watson, M.S.; Perry, J.; Cullinane, A.A.; Davison, A.J. The DNA Sequence of Equine Herpesvirus-4. *DNA Seq.* **1998**, *79*, 1197–1203. [CrossRef]
44. Telford, E.A.R.; Watson, M.S.; McBride, K.; Davison, A.J. The DNA Sequence of Equine Herpesvirus-1. *Virology* **1992**, *189*, 304–316. [CrossRef]

 *animals*

Article

Serological Evidence of Common Equine Viral Infections in a Semi-Isolated, Unvaccinated Population of Hucul Horses

Barbara Bażanów [1,*], Janusz T. Pawęska [2], Aleksandra Pogorzelska [1], Magdalena Florek [1], Agnieszka Frącka [3], Tomasz Gębarowski [4], Wojciech Chwirot [5] and Dominika Stygar [6]

1 Division of Microbiology, Department of Pathology, Faculty of Veterinary Medicine, Wroclaw University of Environmental and Life Sciences, 50-375 Wroclaw, Poland; aleksandra.pogorzelska@upwr.edu.pl (A.P.); magdalena.florek@upwr.edu.pl (M.F.)

2 Centre for Emerging Zoonotic and Parasitic Diseases, National Institute for Communicable Diseases, Johannesburg 2131, South Africa; januszp@nicd.ac.za

3 Grove Referrals, Fakenham NR21 8JG, UK; agnieszkafracka@gmail.com

4 Department of Basic Medical Sciences, Faculty of Pharmacy with the Division of Laboratory Diagnostics, Wroclaw Medical University, 50-556 Wroclaw, Poland; tomasz.gebarowski@umed.wroc.pl

5 Faculty of Veterinary Medicine, Wroclaw University of Environmental and Life Sciences, 50-375 Wroclaw, Poland; 106377@student.upwr.edu.pl

6 Chair and Department of Physiology, School of Medicine with Dentistry Division in Zabrze, Medical University of Silesia, 40-055 Katowice, Poland; dstygar@sum.edu.pl

* Correspondence: barbara.bazanow@upwr.edu.pl

Citation: Bażanów, B.; Pawęska, J.T.; Pogorzelska, A.; Florek, M.; Frącka, A.; Gębarowski, T.; Chwirot, W.; Stygar, D. Serological Evidence of Common Equine Viral Infections in a Semi-Isolated, Unvaccinated Population of Hucul Horses. *Animals* **2021**, *11*, 2261. https://doi.org/10.3390/ani11082261

Academic Editors: Romain Paillot and Thomas W. Swerczek

Received: 17 June 2021
Accepted: 29 July 2021
Published: 30 July 2021

Publisher's Note: MDPI stays neutral with regard to jurisdictional claims in published maps and institutional affiliations.

Simple Summary: The aim of the following research was an analysis of the occurrence of common equine viral infections in a Hucul herd, based on serological studies. This study provides epidemiological data concerning animals representing a semi-isolated herd, devoid of any specific prophylaxis. The obtained results provide a source of information on a primitive breed of horses, as well as giving an insight into viruses (among them, arboviruses) circulating within a local ecosystem.

Abstract: Huculs (*Equus caballus*) are an old breed of primitive mountain horses, originating from the Carpathian Mountains. To the best of our knowledge, data concerning the epidemiology of viral infections observed within this breed are sparse. The objective of this study was to estimate the serological status of a semi-isolated, unvaccinated Hucul herd, with respect to both common equine viral infections and horse-infecting arboviruses, the presence of which was previously reported in Poland. Twenty horses of the Hucul breed, living in a remote area in Poland, were studied in 2018 from March to May. Using nasal secretion swabs as a specimen source, isolation attempts were negative regarding ERAV, EHV-1, EAV, and EIV. According to the virus neutralisation method, in the sera obtained from the animals, antibodies against the following viruses were detected: EHV-1 in 12 horses (60%; with titres from 1:8 to 1:64), EIV A/H7N7 in 13 (65%; titres from 1:20 to 1:80), EIV A /H3N8 in 12 (60%; titres from 1:20 to 1:80), USUV in 5 (25%; titres from 1:10 to 1:80), and ERAV in 1 (5%; titre 1:32). Antibodies against EAV, EIAV, and WNV were not present in the tested sera. The detected presence of specific antibodies associated with five out of the eight equine viruses investigated indicates that the Hucul herd, due to its partial separation and lack of specific prophylaxis, could serve as a sentinel animal group for the detection of equine viruses/arboviruses present within the local ecosystem. The detection of common equine viral infections within the herd provides additional epidemiological data concerning the breed.

Keywords: Huculs; viral status; immunological status; equine viral diseases

1. Introduction

Huculs (*Equus caballus*), also known as Carpathian horses or Carpathian ponies, are a breed of primitive, small, mountain horses [1]. This old breed originates from the region of the eastern part of the Carpathian Mountains, and the first written material concerning these animals dates back to the beginning of the 17th century [1,2]. It is assumed that Huculs may be descendants of Tarpan horses mixed with Arabian horses, as well as Lipican, Hafling, Norik breed, and Fordwhich [2]. Since the Hucul horse is classified as a small-numbered breed, it has been a part of the FAO Program for the Preservation of Animal Genetic Resources [3]. Huculs are distinguished by their incredible strength, vitality, and resistance to harsh weather conditions; therefore, they can be kept on pastures throughout the year [1]. They are also known for their excellent health, including disease resistance, as well as for their high fertility and longevity [3,4]. The data concerning this breed are sparse (28 records in PubMed). To the best of our knowledge, epidemiological information in the form of only two papers exclusively reporting equine viral arteritis (EVA) infections is available for these animals [5,6].

The most common viral infections observed in equids are caused by viral agents belonging to different taxonomic families, including equine arteritis virus (EAV, *Arteriviridae*), equine herpesvirus 1 (EHV-1, *Herpesviridae*), equine rhinitis A virus (ERAV, *Picornaviridae*), and equine influenza A virus (EIV, *Orthomyxoviridae*) [7,8]. Additionally, when it comes to these animals, infections caused by Usutu virus (USUV, *Flaviviridae*), West Nile virus (WNV, *Flaviviridae*), and equine infectious anaemia virus (EIAV, *Retroviridae*) are of recently increasing importance [9,10].

Equine viral arteritis (EVA), caused by EAV, is a global infectious disease of horses characterised by abortion in pregnant mares. Other clinical signs like anorexia, depression, conjunctivitis, respiratory signs, and ocular discharge, as well as oedema of the eyelids, abdomen, prepuce and scrotum, or mammary glands, can also be observed [7,11]. Infection with EHV-1 is highly prevalent worldwide. Whereas it may be inapparent, in some animals it results in abortion or presents as acute rhinitis and pharyngitis, with the potential to spread into the more distal airways, leading to tracheobronchitis, bronchiolitis, and/or pneumonitis [7,12,13]. Other infections of the upper respiratory tract commonly observed in equids are caused by ERAV and the two main strains of EIV: equine-1 (H7N7) and equine-2 (H3N8) influenza A virus [14,15]. It is worth mentioning that the ERAV may also infect humans [16]. USUV and WNV are mosquito-borne zoonotic arboviruses transmitted by *Culex pipiens*, causing severe encephalitis and death in horses, birds, and humans [17]. The presence of antibodies to these two arboviruses, which were originally bound to Africa or Southeastern Asia, was previously confirmed in Poland in both humans and animals [9,18]. Bloodsucking insects, especially horseflies and deerflies, are EIAV vectors. In horses, symptoms of infections caused by the EIAV include high or recurrent fever, anaemia, weakness, and swelling of the lower abdomen, chest, legs, and scrotum, as well as weak pulse, irregular heartbeat, and abortion in pregnant mares. The latter disease presents no public health risk [19].

It was established that animal populations can serve as sentinels for the detection of environmental health threats [20]. According to the Centers for Disease Control and Prevention, the aim of sentinel surveillance is to obtain timely information in a relatively inexpensive manner, rather than to derive precise estimates of prevalence or incidence in the general population [20].

The objective of this study was to analyse the virological and immunological status of a herd of Hucul horses living in semi-natural conditions (and not vaccinated or artificially inseminated) with respect to common equine viral infections and horse-infecting arboviruses, the presence of which was previously reported in Poland. The method of maintenance and lack of medical intervention makes the studied herd, despite its small size, a useful model of sentinel animals/surveillance.

2. Materials and Methods

2.1. Ethics Statement

The blood samples and nasal swabs used in this project were taken for the routine diagnosis of EAV infections. The methods for the specimens' collection were carried out in accordance with article 37ah–37ak of the Pharmaceutical Law act (Dz. U. z 2015 r. poz. 266, zmienione przez Dz. U. z 2019 r. poz. 499, 399 i 959 [Journal of Laws of the Republic of Poland from 2015, item 266 changed with Journal of Laws of the Republic of Poland from 2019, item. 499, 399 and 959]) from September 6th 2001 and the Experiments on Animals Act (Dz. U. 2019 poz. 1392 [Journal of Laws of the Republic of Poland 2019, item. 1392]) from 5 July 2019. In light of these regulations, approval of the Ethical Committee was not required. Informed consent was obtained from the owner of the animals.

2.2. Specimen

A herd of 20 Hucul horses in total (10 mares, 10 stallions) aged between 5 and 10 years, located in southwestern Poland (in Kalisz District, Greater Poland Voivodeship), was tested from March to May 2018. The animals had no contact with either other horses or wild equines, and they were kept on extensive pasture distant from inhabited areas throughout the year. None of the horses showed clinical symptoms at the time of sampling. There was no sign of clinical infection reported during the year prior to sampling, except for two incidents of abortion (for non-infectious reasons) and a single case of mild respiratory symptoms that was not diagnosed. All the horses were therefore considered healthy on the basis of physical examination and haematological and biochemical analyses (data not shown). None of the Huculs were vaccinated against any equine viral disease.

For serological tests, blood samples were collected from all the horses. Blood specimens were allowed to clot and were then centrifuged at 3000× *g* for 10 min, then the obtained sera were transferred into fresh tubes. Nasal swabs taken from the animals were transported to the laboratory in tubes containing EMEM supplemented with antibiotics, at 4 °C. The swabs were processed by pressing against the tube walls and vortexing, using the transport medium. The obtained fluid was centrifuged for 10 min at 3000× *g*. The secured supernatants were transferred into fresh tubes.

2.3. Virus Isolation

Attempted virus isolation of EAV, EHV-1, and ERAV from nasal swabs was performed using standard isolation procedures. Twenty-four-well polystyrene plates containing rabbit kidney (RK-13, ATCC® CCL-37™, ATCC, Manassas, VA, USA) or green monkey kidney (Vero, ATCC® CCL-81™) cell lines were inoculated with the processed swab specimens (50–100 µL into each well) [21]. Plates were incubated at 37 °C/5% CO_2 and observed daily for 7–10 days, for the development of a cytopathic effect (CPE), using an inverted microscope (Olympus Corp., Hamburg Germany; Axio Observer, Carl Zeiss Microscopy Deutschland GmbH, Jena, Germany). In the absence of visible CPE, up to 5 blind subsequent passages were performed. Attempted virus isolation of EIV was carried out in embryonated chicken eggs (ECEs) [22]. The samples from nasal swabs were inoculated into the allantoic cavities of the ECEs. After inoculation, the ECEs were kept for 4–5 days in an incubator for chicken eggs at 38 °C and then cooled at 4 °C for 24 h, opened, and examined for the presence of any changes in the embryos and on the membranes. Allantoic fluid was harvested and tested by hemagglutination assay (HA) using red blood cells from chickens (RBC aggregation was regarded as a positive result).

2.4. Serology

Sera obtained from all the horses were tested for the presence of antibodies against EAV, EHV-1, ERAV, EIV (H7N7 and H3N8), USUV, WNV, and EIAV. Virus neutralisation (VN) tests were performed using the Bucyrus, Rac-H, and V1722/70 reference strains for the detection of EAV [23], EHV1 [24], and ERAV [25], respectively. In order to detect antibodies to EIV, a haemagglutination inhibition test (HI) was carried out according to the World Health Organisation (WHO) recommendations [26] using A/equine/Miami/1/63 (H3N8) and A/equine/Prague/1/56 (H7N7) reference strains.

A microneutralisation assay was applied for the detection of antibodies against USUV and WNV, using the previously described procedure [9,27]. The presence of precipitating antibodies against the EIAV gag p26 protein [28,29] was tested via the agar gel immunodiffusion test (AGID), employing a commercial Equine Infectious Anemia Antibody Test Kit LAB-EZ /EIA (Zoetis, Parsippany, NJ, USA).

All the VN and HI tests were carried out in the presence of control sera (positive and negative) from the in-house collection.

2.5. Statistical Analysis

Statistical analysis was conducted in PQStat version 1.6.8 (PQ Stat Software, Poznań, Poland) at a significance level of 5% using Fisher's test. The correlation between the age and seropositivity to individual pathogens was evaluated. For this purpose, the animals were divided by age into two groups: 5 years old (group of youngest horses in the herd) and over 5 years old.

3. Results

According to the results of the virus isolation, all the nasal swabs were negative for the presence of EAV, EHV-1, ERAV, and EIV. Specific antibodies were detected for five out of the eight examined equine viruses. Detailed results of our serological testing are shown in Tables 1 and 2.

Table 1. Results of serological investigation of the Hucul herd.

Virus Tested	Number of Positive Results/Number of the Horses Tested	Percentage of Seropositive Horses (%)
equine arteritis virus	0/20	0
equine herpes virus 1	12/20	60
equine rhinitis A virus	1/20	5
equine influenza A virus H7N7	13/20	65
equine influenza A virus H3N8	12/20	60
Usutu virus	5/20	25
West Nile virus	0/20	0
equine infectious anaemia virus	0/20	0

Table 2. The immune status in individual horses.

No.	Sex	Age (Years)	Obtained Titres of Antibodies							
			EAV	EHV1	ERAV	H7N7	H3N8	USUV	WNV	EIAV
1	mare	10	0	0	0	40	20	40	0	0
2	mare	9	0	16	0	0	20	80	0	0
3	mare	5	0	0	0	20	0	20	0	0
4	mare	10	0	32	8	0	0	20	0	0
5	mare	5	0	0	0	20	0	0	0	0
6	mare	8	0	0	0	0	80	0	0	0
7	mare	10	0	0	0	20	20	0	0	0
8	mare	5	0	8	0	40	40	0	0	0
9	mare	5	0	32	0	40	40	0	0	0
10	mare	9	0	32	0	0	40	0	0	0
11	stallion	5	0	0	0	80	40	0	0	0
12	stallion	6	0	32	0	20	0	10	0	0
13	stallion	6	0	8	0	20	20	0	0	0

Table 2. Cont.

No.	Sex	Age (Years)	Obtained Titres of Antibodies							
			EAV	EHV1	ERAV	H7N7	H3N8	USUV	WNV	EIAV
14	stallion	5	0	32	0	0	0	0	0	0
15	stallion	5	0	16	0	20	0	0	0	0
16	stallion	5	0	0	0	0	0	0	0	0
17	stallion	9	0	16	0	20	20	0	0	0
18	stallion	10	0	8	0	0	20	0	0	0
19	stallion	9	0	16	0	20	0	0	0	0
20	stallion	5	0	0	0	20	20	0	0	0

Of the 20 Huculs tested, 12 (60%) were positive for antibodies to EHV-1 with titres ranging from 1:8 to 1:32. Antibodies to equine influenza viruses A/H7N7 and A /H3N8 were found in the sera of 13 (65%) and 12 (60%) horses, respectively, with titres ranging from 1:20 to 1:80 against both viruses. USUV seropositivity was detected in five horses (25%) with titres ranging from 1:10 to 1:80. Only one (5%) animal had antibodies to ERAV (titre of 1:8). None of the horses had detectable antibody levels for EAV, EIAV, or WNV.

Statistical analysis showed no significant correlations between the variables of age group and antibody presence for viruses EHV1 ($p > 0.999$), ERAV ($p > 0.999$), H7N7 ($p = 0.3498$), H3N8 ($p = 0.1698$), and USUV ($p > 0.999$). Due to the negative results of the serological investigation, statistical analyses for age vs. antibody presence for EAV, WNV, and EIAV were not performed.

4. Discussion

Nasal swabs obtained from all the horses tested in this study were negative for the presence of the four common equine viruses investigated, which indicates the absence of ongoing infection with EAV, EHV-1, ERAV, or EIV at the time of swabbing.

Serum neutralising antibodies following infection with EAV have been shown to persist for years, and it was postulated that humoral immunity to this virus is long-lasting, if not lifelong [5,7]. Thus, the absence of detectable anti-EAV antibodies in the horses tested in our study indicates that they had no previous exposure to the virus, for which the main route of infection is direct contact (although spread through fomites has also been documented) [6]. Compared with our results, analysis of the largest Hucul stud in Poland (where the animals are commercially used), performed by Rola et al. in the years 2006–2008, showed the presence of antibodies to EAV in 55.1% of the horses [5]. Subsequent analysis concerning the same stud, carried out between the years 2010 and 2013, confirmed the presence of anti-EAV antibodies in 38% of the collected samples [6]. With respect to the total equid population in Poland, antibodies against EAV were detected in around 20% of the tested horses [30], and EVA was identified as a cause of 23% of abortions in mares [31]. The EAV seroprevalence reported in Europe is highly diverse. The lowest ratio was observed in the U.K. (1.3%). Reference countries in western and central Europe presented a range of seroprevalence from 14% to 20%, while in eastern Europe, it oscillated within the range of 14–75% [32–37].

EHV-1 is highly prevalent, spreads quickly among horses, and establishes a latent carrier state, which may possibly last for life. In latently infected animals, stressors, e.g., transport or pregnancy, can induce respiratory viral shedding [12,13]. The percentage of horses seropositive for EHV-1 in the population of southeastern Poland's breeding farms, stallion herds, purchasing centres, and riding clubs ranges from 75% to 100% [38]. Compared to the situation in Poland, in Spain, only 53.9% of horses tested seropositive for EHV-1, while in the Netherlands, this proportion was even smaller, as 28.2% of the animals were positive [39,40]. In the present study, antibodies against EHV-1 were detected in 60% of the Huculs tested. Taking into consideration the facts that the spread of the virus requires direct contact and clinical symptoms suggesting viral infection were not observed within the herd, we assume that the virus might have been introduced by a previously latently infected animal.

Antibodies to ERAV were detected in only 5% of the horses tested. This relatively low level of ERAV infections is probably the result of the separation of the tested herd. In Poland, high ERAV seropositivity (72%) was reported in different breeds of horses [8]. In England, Switzerland, and Denmark, the observed ERAV seropositivity levels ranged from 2.3% to 14%; contrastingly, research showed that in Germany and the Netherlands, these values were far higher (39% to 90%) [14,16,41,42].

In the present study, a high percentage of EIV seropositivity was detected against subtypes H7N7 (65%) and H3N8 (60%). Since the first isolation in 1963 in the USA, the subtype H3N8 (Miami 63) of EIV has continued to circulate among horses, causing epidemics and minor outbreaks of equine influenza respiratory disease in Europe and North America [15,43,44]. The last major outbreak caused by H3N8 in Europe took place in 2018/2019 and involved the U.K., Ireland, France, Germany, and Belgium. While in Europe, the most commonly found is Florida clade 2, the recently reported EIV infections were caused by Florida clade 1, which had, until then, been circulating mainly in the United States [45,46].

The EIV H7N7 subtype (Prague 56) was first isolated in 1956 from horses in Eastern Europe [47]. This subtype has not been isolated since the last outbreak recorded in 1979 and was thought to be extinct. However, antibodies to this subtype are still detectable in horses, suggesting that it still circulates in equine populations [44]. During the last decade of the 20th century, H7N7-positive horses were reported in Belgium, Russia, the USA, India, and Nigeria [44,48,49]. Interestingly, in their study, Guo et al. reported that there was cross-reactivity observed within the H7 subtype influenza virus. The cross-reactivity was detected between H7N9 and subtypic viruses H7N2, H7N3, and H7N7 [50]. An analysis of khulan (*Equus hemionus hemionus*) in Mongolia, performed in 2015, showed that two of the tested animals were positive for the H7 virus type. While cross-reaction with the H7N9 strain was excluded in both cases, antibodies obtained from the horses were assigned as H7N7 using a protein microarray (PA) technique, but when one of the sera was additionally tested via single radial haemolysis assay, it proved to be H7N3 positive. The authors suggested co-circulation of both subtypes, which are of equine and avian origin, respectively [51]. Antibodies to the H7 protein were also detected in the same animal species in Nigeria [52], where a haemagglutination inhibition technique using H7N3 antigen was applied. Taking into the consideration the results of the above-mentioned investigations, the positive sera obtained in our study may represent remains of the circulating H7N7 equine-originating virus or the presence of other H7 subtypes, most probably of avian origin. The number of animals seropositive to both EI viruses among the Huculs tested in the present study was surprisingly high, particularly regarding the fact that these horses were not vaccinated against equine influenza and were living in isolation, without contact with other domestic animals. In the late 1980s in China, an outbreak caused by the H3N8 virus strain more closely related to avian H3 than to the equine H3 influenza virus was described [48]. Moreover, analysis of the outbreak of EI in New South Wales in 2007 suggested the potential role of birds as mechanical spreaders of the equine H3N8 virus [53]. Both scenarios equally explain the results of our investigation, yet other mechanisms of transmission of the EI viruses to the herd tested in the present study cannot be excluded.

In Europe, the USUV was isolated for the first time from bats in Germany, and the first human infection with this virus took place in Italy in 2009; three years later, in Germany, research showed that 1 out of 4200 patients had antibodies against USUV [18,54,55]. The relatively high USUV seropositivity detected in the present study corroborates the results of a survey recently carried out in Poland, where 27.98% of the tested horses were seropositive to this virus [9]. However, these results are quite different from the findings of a study conducted in Croatia, where USUV-specific neutralising antibodies were detected in 2 of the total 1380 animals [56]. The seroprevalence of USUV in European horses is highly varied. In Serbia, it was found to be only 0.3%, and in Spain, it was 1.4%; however, in Italy it was as high as 89.2% [57]

The first report of the appearance of WNV in Poland comes from the years 1995–1996, when Juricova et al. conducted research on two sparrow species and stated a relatively high prevalence (2.8% in house sparrow (*Passer domesticus*) and 12.1% in Eurasian tree sparrow (*Passer montanus*)) of anti-WNV antibodies in the sera of the tested birds [58]. In 2013 in the Czech Republic, Rudolf et al., while testing mosquitoes, found WNV strains that were genetically similar to those isolated in Austria, Italy, and Serbia, which may testify to the spreading of this virus across all of Europe [18]. In studies performed in Poland, antibodies to WNV were reported in 35.7% of avian sera [9]. Two serological surveys concerning the prevalence of WNV in horses in Poland showed the presence of the antibodies in 0.65% of sera collected before 2008 and in 15.08% of samples obtained in the years 2012–2013 [9,30]. On the contrary, all the Huculs examined in the present study were negative for anti-WNV antibody presence. WNV infections have also been confirmed in various European countries like Serbia, Croatia, Greece, Slovakia, Spain, and Albania, and the seroprevalence of this virus in horses was found to range from 2.75% to 22% [59].

The absence of detectable levels of antibodies to EIAV observed in the present study was an expected result, considering the fact that since the last confirmed case of EIAV in Poland was in 1960, and only two additional cases were reported in 2015 [60]. Between 2007 and 2014, however, outbreaks of the disease were reported in Belgium, Bosnia, Croatia, France, Germany, Greece, Hungary, Ireland, Italy, Latvia, Romania, Serbia, Slovenia, and the United Kingdom [10], indicating a need for ongoing monitoring of the disease's spread.

The serological data obtained for the herein-analysed small group of Hucul horses, especially information concerning the absence of antibodies against certain viruses (EAV, EIAV, and WNV), cannot be extrapolated with regard to the whole Hucul population in Poland (over 1500 animals in total [2]). In order to understand the epidemiological status of the breed in our country in a statistically appropriate manner, a larger number of analyses need to be performed.

5. Conclusions

The present study provides useful epidemiological data regarding viral infections in Hucul horses, supplementing the sparse information on the matter available to date. Partial isolation of the tested herd, in addition to the fact that it was also devoid of any specific prophylaxis, allowed us to analyse the circulation within a local ecosystem of viruses causing infections in horses. Among these investigated viruses were those transmitted via vectors. The fact that the tested animals were apparently healthy at the moment of sampling and throughout the year prior suggests that the detected antibodies are a result of earlier exposures, but it may also be related to the natural resistance of Huculs, as indicated by other authors [1,5]. In such a case, in order to recognise the epidemiology of viral infections concerning the breed, further serological surveys are required.

Author Contributions: Conceptualisation, B.B., M.F. and D.S.; Data analysis, B.F. and D.S.; Investigation, B.B., A.F., J.T.P., A.P., T.G., W.C. and D.S.; Methodology, B.B., J.T.P.; Supervision, B.B.; Writing—original draft, B.B., M.F., A.F., A.P., T.G., J.T.P. and D.S. All authors have read and agreed to the published version of the manuscript.

Funding: The research is financed/co-financed under the Leading Reasarch Groups support project from the subsidy increased for the period 2020–2025 in the amount of 2% of the subsidy referred to Art. 387 (3) of the Law of 20 July 2018 on Higher Education and Science, obtained in 2019.

Institutional Review Board Statement: Ethical review and approval were waived for this study, due to the country's current laws (see the text for details, in the "Materials and Methods" paragraph).

Data Availability Statement: The original data are available by contact with the corresponding author.

Acknowledgments: The authors would like to express their special gratitude to James Tattersall and Scott Richards for scientific English language correction.

Conflicts of Interest: The authors declare that they have no conflict of interest.

References

1. Purzyc, H.A. General characteristic of Hucul horses. *Act. Sci. Pol. Med. Vet.* **2000**, *6*, 25–31.
2. Mackowski, M.; Mucha, S.; Cholewiński, G.; Cieslak, J. Genetic diversity in Hucul and Polish primitive horse breeds. *Arch. Anim. Breed.* **2015**, *58*, 23–31. [CrossRef]
3. Cywińska, A.; Czopowicz, M.; Witkowski, L.; Górecka, R.; Degórski, A.; Guzera, M.; Szczubełek, P.; Turło, A.; Schollenberger, A.; Winnicka, A. Reference intervals for selected hematological and biochemical variables in Hucul horses. *Pol. J. Vet. Sci.* **2015**, *18*, 439–445. [CrossRef]
4. Fornal, A.; Radko, A.; Piestrzyńska-Kajtoch, A. Genetic polymorphism of Hucul horse population based on 17 microsatellite loci. *Act. Bioch. Pol.* **2013**, *60*, 761–765. [CrossRef]
5. Rola, J.; Larska, M.; Rola, J.G.; Belák, S.; Autorino, G.L. Epizootiology and phylogeny of equine arteritis virus in hucul horses. *Vet. Microbiol.* **2011**, *148*, 402–407. [CrossRef]
6. Socha, W.; Sztromwasser, P.; Dunowska, M.; Jaklińska, B.; Rola, J. Spread of equine arteritis virus among Hucul horses with different EqCXCL16 genotypes and analysis of viral quasispecies from semen of selected stallions. *Sci. Rep.* **2020**, *10*, 290. [CrossRef] [PubMed]
7. Bażanów, B.A.; Jackulak, N.A.; Frącka, A.B.; Staroniewicz, Z.M. Abortogenic viruses in horses. *Equine Vet. Educ.* **2014**, *26*, 48–55. [CrossRef]
8. Bażanów, B.A.; Frącka, A.B.; Jackulak, N.A.; Romuk, E.; Gębarowski, T.; Owczarek, A.; Stygar, D. Viral, Serological, and Antioxidant Investigations of Equine Rhinitis A Virus in Serum and Nasal Swabs of Commercially Used Horses in Poland. *BioMed Res. Int.* **2018**, 1–6. [CrossRef]
9. Bażanów, B.A.; Jansen van Vuren, P.; Szymański, P.; Stygar, D.; Frącka, A.; Twardoń, J.; Kozdrowski, R.; Pawęska, J.T. A Survey on West Nile and Usutu Viruses in Horses and Birds in Poland. *Viruses* **2018**, *10*, 87. [CrossRef] [PubMed]
10. Bolfa, P.; Barbuceanu, F.; Leau, S.E.; Leroux, C. Equine infectious anaemia in Europe: Time to re-examine the efficacy of monitoring and control protocols. *Equine Vet. J.* **2016**, *48*, 140–142. [CrossRef] [PubMed]
11. Gerber, H.; Steck, F.; Hofer, B.; Walther, L.; Friedli, U. Serological investigations on equine viral arteritis. In Proceedings of the Fourth International Conference on Equine Infectious Diseases, Lyon, France, 24–27 September 1976; Bryans, J.T., Gerber, H., Eds.; Veterinary Publications: Princeton, NJ, USA, 1978; pp. 461–465.
12. Gilkerson, J.R.; Whalley, J.M.; Drummer, H.E.; Studdert, M.J.; Love, D.N. Epidemiology of EHV-1 and EHV-4 in the mare and foal populations on a Hunter Valley stud farm: Are mares the source of EHV-1 for unweaned foals. *Vet. Microbiol.* **1999**, *68*, 27–34. [CrossRef]
13. Bażanów, B.A.; Jackulak, N.A.; Florek, M.; Staroniewicz, Z. Equid Herpesvirus-Associated Abortion in Poland between 1977–2010. *J. Equine Vet. Sci.* **2012**, *32*, 747–751. [CrossRef]
14. Black, W.; Wilcox, R.; Stevenson, R.; Hartley, C.; Ficorilli, N.; Gilkerson, J.; Studdert, M. Prevalence of serum neutralising antibody to equine rhinitis A virus (ERAV), equine rhinitis B virus 1 (ERBV1) and ERBV2. *Vet. Microbiol.* **2007**, *119*, 65–71. [CrossRef] [PubMed]
15. Burrows, R.; Denyer, M.; Goodridge, D.; Hamilton, F. Field and laboratory studies of equine influenza viruses isolated in 1979. *Vet. Rec.* **1981**, *109*, 353–356. [CrossRef] [PubMed]
16. Kriegshäuser, G.; Deutz, A.; Kuechler, E.; Skern, T.; Lussy, H.; Nowotny, N. Prevalence of neutralizing antibodies to Equine rhinitis A and B virus in horses and man. *Vet. Microbiol.* **2005**, *106*, 293–296. [CrossRef] [PubMed]
17. Nikolay, B. A review of West Nile and Usutu virus co-circulation in Europe: How much do transmission cycles overlap? *Trans. R. Soc. Trop. Med. Hyg.* **2015**, *109*, 609. [CrossRef] [PubMed]
18. Moniuszko-Malinowska, A.; Czupryna, P.; Dunaj, J.; Zajkowska, J.; Siemieniako, A.; Pancewicz, S. West Nile virus and USUTU–a threat to Poland. *Przegl. Epidemiol.* **2016**, *70*, 7–10. [PubMed]
19. Issel, C.J.; Adams, W.V. Detection of equine infectious anemia virus in a horse with an equivocal agar gel immunodiffusion test reaction. *J. Am. Vet. Med. Assoc.* **1982**, *180*, 276–278. [PubMed]
20. Halliday, J.E.B.; Meredith, A.L.; Knobel, D.L.; Shaw, D.J.; Bronsvoort, B.M.d.C.; Cleaveland, S. A framework for evaluating animals as sentinels for infectious disease surveillance. *J. R. Soc. Interface* **2007**, *4*, 973–984. [CrossRef]
21. Timoney, J.; Gillespie, J.H.; Scott, F.W.; Barlough, J.E. Laboratory diagnosis of viral infection. In *Hagan and Bruner's Microbiology and Infectious Diseases of Domestic Animals*, 8th ed.; Cornell University Press: Ithaca, NY, USA, 1988; pp. 467–468.
22. Cullinane, A. Equine influenza (infection with equine influenza virus). In *Manual of Diagnostic Tests and Vaccines for Terrestrial Animals 2019*, 8th ed.; OIE World Organisation for Animal Health: Paris, France, 2020; pp. 1–19.
23. Timoney, P.J. Equine viral arteritis (infection with equine arteritis virus). In *Manual of Diagnostic Tests and Vaccines for Terrestrial Animals 2019*, 8th ed.; OIE World Organisation for Animal Health: Paris, France, 2020; pp. 1333–1349.
24. Elton, D.; Bryant, N. Equine rhinopneumonitis (equine herpesvirus-1 and -4). In *Manual of Diagnostic Tests and Vaccines for Terrestrial Animals 2019*, 8th ed.; OIE World Organisation for Animal Health: Paris, France, 2020; pp. 1320–1332.

25. Carman, S.; Rosendal, S.; Huber, L.; Gyles, C.; McKee, S.; Willoughby, R.A.; Dubovi, E.; Thorsen, J.; Lein, D. Infecticus agents in acute respiratory disease in horses in Ontario. *J. Vet. Diagn. Investig.* **1997**, *9*, 17–23. [CrossRef]
26. Zhang, W.; Besselaar, T.G. The laboratory diagnosis and virological surveillance of influenza. In *Manual for the Laboratory Diagnosis and Virological Surveillance of Influenza*; World Health Organization: Geneva, Switzerland, 2011; pp. 59–63.
27. Ludolfs, D.; Niedrig, M.; Paweska, J.T.; Schmitz, H. Reverse ELISA for the detection of anti-West Nile virus antibodies in humans. *Eur. J. Clin. Microbiol. Infect. Dis.* **2007**, *26*, 467–473. [CrossRef]
28. Oliveira, F.G.; Diniz, R.S.; Camargos, M.F.; de Oliveira, A.M.; de Souza Rajão, D.; Braz, G.F.; Leite, R.C.; Pimenta dos Reis, J.K. Comparative study of agar gel immunodiffusion (AGID) protocols for the diagnosis of equine infectious anemia in Brazil. *Sem. Cien. Agra.* **2013**, *34*, 3909–3916. [CrossRef]
29. Mealey, R.H. Equine Infectious Anaemia. In *Equine Infectious Diseases*, 1st ed.; Sellon, D.C., Long, M.T., Eds.; Saunders Elsevier: Amsterdam, The Netherlands, 2007; pp. 213–219.
30. Golnik, W.; Bażanów, B.; Florek, M.; Pawęska, J. Prevalence of equine arteritis and West Nile virus–specific antibodies in thoroughbred horses in Poland. *Ann. Univ. Mariae Curie-Sklodowska* **2008**, *21*, 1–4. [CrossRef]
31. Bażanów, B.A.; Frącka, A.B.; Jackulak, N.A.; Staroniewicz, Z.M.; Ploch, S.M. A 34-year retrospective study of equine viral abortion in Poland. *Pol. J. Vet. Sci.* **2014**, *17*, 607–612. [CrossRef]
32. Equine viral arteritis: Not just a reproductive disease. *Vet. Rec.* **2019**, *184*, 791–793. [CrossRef] [PubMed]
33. Cruz-Lopez, F.; Newton, R.; Sanchez-Rodriguez, A.; Ireland, J.; Mughini-Gras, L.; Moreno, M.A.; Fores, P. Equine viral arteritis in breeding and sport horses in central Spain. *Res. Vet. Sci.* **2017**, *115*, 88–91. [CrossRef] [PubMed]
34. Lazić, S.; Lupulović, D.; Gaudaire, D.; Petrovic, T.; Lazić, G.; Hans, A. Serological evidence of equine arteritis virus infection and phylogenetic analysis of viral isolates in semen of stallions from Serbia. *BMC Vet. Res.* **2017**, *13*, 316. [CrossRef] [PubMed]
35. Newton, J.R.; Wood, J.L.; Castillo-Olivares, F.J.; Mumford, J.A. Serological surveillance of equine viral arteritis in the United Kingdom since the outbreak in 1993. *Vet. Rec.* **1999**, *145*, 511–516. [CrossRef]
36. Kölbl, S.; Schuller, W.; Pabst, J. Serologische Untersuchungen zur aktuellen Verseuchung österreichischer Pferde mit dem Virus der Equinen Arteritis [Serological studies of the recent infections of Austrian horses with the equine arteritis virus]. *Dtsch. Tierarztl. Wochenschr.* **1991**, *98*, 43–45. (In German)
37. Szeredi, L.; Hornyák, A.; Pálfi, V.; Molnár, T.; Glávits, R.; Dénes, B. Study on the epidemiology of equine arteritis virus infection with different diagnostic techniques by investigating 96 cases of equine abortion in Hungary. *Vet. Microbiol.* **2005**, *108*, 235–242. [CrossRef]
38. Grądzki, Z.; Boguta, L. Seroprevalence of EHV1 and EHV4 in the horse population of the southeastern part of Poland. *Med. Wet.* **2009**, *65*, 188–193.
39. Patel, J.R.; Heldens, J. Equine herpesviruses 1 (EHV-1) and 4 (EHV-4)—Epidemiology, disease and immunoprophylaxis: A brief review. *Vet. J.* **2005**, *170*, 14–23. [CrossRef] [PubMed]
40. Cruz, F.; Fores, P.; Mughini-Gras, L.; Ireland, J.; Moreno, M.A.; Newton, J.R. Seroprevalence and factors associated with equine herpesvirus type 1 and 4 in Spanish Purebred horses in Spain. *Vet. Rec.* **2016**, *178*, 398. [CrossRef] [PubMed]
41. Holmes, D.F.; Kemen, M.J.; Coggins, L. Equine rhinovirus infection—Serologic evidence of infection in selected horse population. In *Equine Infectious Diseases*, 4th ed.; Bryans, J.T., Gerber, H., Eds.; Karger: Basel, Switzerland, 1978; pp. 315–319.
42. De Boer, G.F.; Osterhaus, A.D.; van Oirschot, J.T.; Wemmenhove, R. Prevalence of antibodies to equine viruses in the Netherlands. *Vet. Q.* **1979**, *1*, 65–74. [CrossRef] [PubMed]
43. Daly, J.M.; Yates, P.J.; Newton, J.R.; Park, A.; Henley, W.; Wood, J.L.N.; Davis-Poynter, N.; Mumford, J.A. Evidence supporting the inclusion of strains from each of the two co-circulating lineages of H3N8 equine influenza virus in vaccines. *Vaccine* **2004**, *22*, 4101–4109. [CrossRef]
44. Webster, R.G. Are equine 1 influenza viruses still present in horses? *Equine Vet. J.* **1993**, *25*, 537–538. [CrossRef]
45. Paillot, R.; Pitel, P.H.; Pronost, S.; Legrand, L.; Fougerolle, S.; Jourdan, M.; Marcillaud-Pitel, C. Florida clade 1 equine influenza virus in France. *Vet. Rec.* **2019**, *184*, 101. [CrossRef]
46. Nemoto, M.; Ohta, M.; Yamanaka, T.; Kambayashi, Y.; Bannai, H.; Tsujimura, K.; Yamayoshi, S.; Kawaoka, Y.; Cullinane, A. Antigenic differences between equine influenza virus vaccine strains and Florida sublineage clade 1 strains isolated in Europe in 2019. *Vet. J.* **2021**, *272*, 105674. [CrossRef]
47. Sovinová, O.; Tumová, B.; Pouska, F.; Nemec, J. Isolation of a virus causing respiratory disease in horses. *Acta Virol.* **1958**, *2*, 51–61.
48. Guo, Y.; Wang, M.; Kawaoka, Y.; Gorman, O.; Ito, T.; Saito, T.; Webster, R.G. Characterisation of a new avian-like influenza A virus from horses in China. *Virology* **1992**, *188*, 245–255. [CrossRef]
49. Adeyefa, C.A.O.; James, M.L.; McCauley, J.W. Antigenic and genetic analysis of equine influenza viruses from tropical Africa in 1991. *Epidemiol. Infect.* **1996**, *117*, 367–374. [CrossRef] [PubMed]
50. Guo, L.; Wang, D.; Zhou, H.; Wu, C.; Gao, X.; Xiao, Y.; Ren, L.; Paranhos-Baccalà, G.; Shu, Y.; Jin, Q.; et al. Cross-reactivity between avian influenza A (H7N9) virus and divergent H7 subtypic- and heterosubtypic influenza A viruses. *Sci. Rep.* **2016**, *6*, 22045. [CrossRef]
51. Soilemetzidou, E.S.; de Bruin, E.; Eschke, K.; Azab, W.; Osterrieder, N.; Czirják, G.Á.; Buuveibaatar, B.; Kaczensky, P.; Koopmans, M.; Walzer, C.; et al. Bearing the brunt: Mongolian khulan (Equus hemionus hemionus) are exposed to multiple influenza A strains. *Vet. Microbiol.* **2020**, *242*, 108605. [CrossRef] [PubMed]

52. Olusa, T.A.O.; Adegunwa, A.K.; Aderonmu, A.A.; Adeyefa, C.A.O. Serologic evidence of equine H7 influenza virus in polo horses in Nigeria. *Sci. World J.* **2010**, *5*, 17–19. [CrossRef]
53. Spokes, P.J.; Marich, A.J.; Musto, J.A.; Ward, K.A.; Craig, A.T.; McAnulty, J.M. Investigation of equine influenza transmission in NSW: Walk, wind or wing? *N. S. W. Public Health Bull.* **2009**, *20*, 152–156. [CrossRef] [PubMed]
54. Cadar, D.; Becker, N.; de Mendonca Campos, R.; Börstler, J.; Jöst, H.; Schmidt-Chanasit, J. Usutu virus in bats, Germany, 2013. *Emerg. Infect. Dis.* **2014**, *20*, 1771–1773. [CrossRef] [PubMed]
55. Allering, L.; Jöst, H.; Emmerich, P.; Günther, S.; Lattwein, E.; Schmidt, M.; Seifried, E.; Sambri, V.; Hourfar, K.; Schmidt-Chanasit, J. Detection of Usutu virus infection in a healthy blood donor from south-west Germany, 2012. *Eurosurveillance* **2012**, *17*. [CrossRef]
56. Barbic, L.; Vilibic-Cavlek, T.; Listes, E.; Stevanovic, V.; Gjenero-Margan, I.; Ljubin-Sternak, S.; Pem-Novosel, I.; Listes, I.; Mlinaric-Galinovic, G.; Di Gennaro, A.; et al. Demonstration of Usutu virus antibodies in horses, Croatia. *Vector-Borne Zoonotic Dis.* **2013**, *13*, 772–774. [CrossRef] [PubMed]
57. Vilibic-Cavlek, T.; Petrovic, T.; Savic, V.; Barbic, L.; Tabain, I.; Stevanovic, V.; Klobucar, A.; Mrzljak, A.; Ilic, M.; Bogdanic, M.; et al. Epidemiology of Usutu Virus: The European Scenario. *Pathogens* **2020**, *9*, 699. [CrossRef]
58. Juricová, Z.; Pinowski, J.; Literák, I.; Hahm, K.H.; Romanowski, J. Antibodies to alphavirus, flavivirus, and bunyavirus arboviruses in house sparrows (Passer domesticus) and tree sparrows (*P. montanus*) in Poland. *Avian. Dis.* **1998**, *42*, 182–185. [CrossRef]
59. Vilibic-Cavlek, T.; Savic, V.; Petrovic, T.; Toplak, I.; Barbic, L.; Petric, D.; Tabain, I.; Hrnjakovic-Cvjetkovic, I.; Bogdanic, M.; Klobucar, A.; et al. Emerging Trends in the Epidemiology of West Nile and Usutu Virus Infections in Southern Europe. *Front. Vet. Sci.* **2019**, *6*, 437. [CrossRef] [PubMed]
60. OIE Equine Infectious Anaemia, Poland. Available online: https://www.oie.int/wahis_2/public/wahid.php/Reviewreport/Review?page_refer=MapFullEventReport&reportid=17935 (accessed on 17 June 2015).

Article

Seroprevalence and Risk Factors for Exposure to Equine Coronavirus in Apparently Healthy Horses in Israel

Gili Schvartz [1,2], Sharon Tirosh-Levy [1], Samantha Barnum [3], Dan David [2], Asaf Sol [2], Nicola Pusterla [3] and Amir Steinman [1,*]

1 Koret School of Veterinary Medicine, The Robert H. Smith Faculty of Agriculture, Food and Environment, The Hebrew University of Jerusalem, Rehovot 7610001, Israel; giliun@gmail.com (G.S.); sharontirosh@gmail.com (S.T.-L.)
2 Division of Virology, Kimron Veterinary Institute, Beit Dagan 50250, Israel; Dand@moag.gov.il (D.D.); Asafs@moag.gov.il (A.S.)
3 Department of Medicine and Epidemiology, School of Veterinary Medicine, University of California, Davis, CA 95616, USA; smmapes@ucdavis.edu (S.B.); npusterla@ucdavis.edu (N.P.)
* Correspondence: amirst@savion.huji.ac.il; Tel.: +972-3-9688-544

Simple Summary: Equine coronavirus (ECoV) is a *β-coronavirus* that, together with other coronaviruses, are pathogenic to both human and animals, as seen in the recent COVID-19 pandemic. ECoV is considered as a diarrheic pathogen in foals and is included in the list of viral causes of enteritis. During the last decade, outbreaks of ECoV were reported in adult horses in the USA, EU and Japan. In Israel, other coronaviruses were reported in cattle, camels and in humans; however, coronaviruses have not been reported in horses. In this study, we aimed to determine the exposure of healthy horses to ECoV and determine the selected risk factors for infection. For this purpose, serum samples were collected from 333 healthy horses, 41 (12.3%) of which had anti-ECoV antibodies. Seropositive horses were found in more than half (58.6%) of the farms and horses located in central Israel were more likely to be positive. ECoV should be included in the differential diagnosis list of pathogens in cases of adult horses with acute onset of anorexia, lethargy, fever and gastrointestinal signs in Israel.

Abstract: Equine coronavirus (ECoV) infection is the cause of an emerging enteric disease of adult horses. Outbreaks have been reported in the USA, EU and Japan, as well as sporadic cases in the UK and Saudi Arabia. Infection of ECoV in horses in Israel has never been reported, and the risk of exposure is unknown. Importation and exportation of horses from and into Israel may have increased the exposure of horses in Israel to ECoV. While the disease is mostly self-limiting, with or without supportive treatment, severe complications may occur in some animals, and healthy carriers may pose a risk of infection to other horses. This study was set to evaluate the risk of exposure to ECoV of horses in Israel by using a previously validated, S1-based enzyme-linked immunosorbent assay (ELISA). A total of 41 out of 333 horses (12.3%) were seropositive. Exposure to ECoV was detected in 17 of 29 farms (58.6%) and the seroprevalence varied between 0 and 37.5% amongst farms. The only factor found to be significantly associated with ECoV exposure in the multivariable model was the geographical area ($p < 0.001$). ECoV should be included in the differential diagnosis list of pathogens in cases of adult horses with anorexia, lethargy, fever and gastrointestinal signs in Israel.

Keywords: equine coronavirus; horse; enteric disease; ECoV; seroprevalence

Citation: Schvartz, G.; Tirosh-Levy, S.; Barnum, S.; David, D.; Sol, A.; Pusterla, N.; Steinman, A. Seroprevalence and Risk Factors for Exposure to Equine Coronavirus in Apparently Healthy Horses in Israel. *Animals* **2021**, *11*, 894. https://doi.org/10.3390/ani11030894

Academic Editor: Harold C. McKenzie III

Received: 24 February 2021
Accepted: 18 March 2021
Published: 21 March 2021

Publisher's Note: MDPI stays neutral with regard to jurisdictional claims in published maps and institutional affiliations.

1. Introduction

Equine coronavirus (ECoV) is a positive single-stranded RNA enveloped virus, belonging to the family *Coronaviridae*. ECoV is a *β-coronavirus*, together with human coronaviruses OC43, 4408 and HKU1, bovine coronavirus (BCoV), porcine hemagglutinating encephalomyelitis virus, canine respiratory coronavirus, and others [1]. Important other members of this group, which are pathogenic to humans, are Severe Acute Respiratory Syndrome Coronavirus (SARS-CoV), Middle-East Respiratory Syndrome Coronavirus (MERS-CoV) and the recent SARS-CoV-2 (the causative agent of COVID-19) [2]. Coronaviruses may cause enteric or respiratory disease in mammalian and avian species [3,4]. ECoV is considered as a diarrheic pathogen in foals and is included in the list of viral causes of enteritis together with rotavirus, adenovirus and parvovirus [5]. In the last decade, ECoV has been associated with outbreaks of enteric disease in adult horses in the United States of America, Europe and Japan [6–11].

Like many other viral diseases, ECoV infection is mostly self-limiting, but supportive care is often required. The most frequent clinical signs pooled from 20 outbreaks in the USA were anorexia (97%), lethargy (88%) and fever (83%) [12]. Possible complications, including necrotizing enteritis, endotoxemia and hyperammonemia-associated encephalopathy, have been reported [4,7,13]. Morbidity rates vary between 10% and 80% [3,10] and mortality rates are usually low [3,10,14], but reached 27% in one outbreak in American miniature horses [15]. Healthy adult horses may shed the virus in the environment via feces, and orofecal route is considered to be the main source of infection to other horses [11,13]. ECoV may be detected by electron microscopy, cell culture isolation and polymerase chain reaction (PCR) [4,11,13]. Serology is based on the detection of antibodies against the ECoV S1 protein using enzyme-linked immunosorbent assay (ELISA) [16,17]. Since 2010, the number of positive cases has steadily increased in the USA, which is probably driven by increased awareness and testing [12]. In the USA, the seroprevalence in healthy horses from 18 states was 9.6%, with the geographic region and specific uses of horses identified as significant risk factors for seropositivity [9]. Prevalence of ECoV in adult horses in other countries is generally unknown. In France, among horses with either respiratory or gastrointestinal signs, 11/395 and 1/200 fecal samples and nasal swabs, respectively, were positive for ECoV by qPCR [6]. In Saudi Arabia and Oman, 5/316 and 0/306 rectal and nasal swabs, respectively, were positive for ECoV by qPCR [4].

To the best of our knowledge, ECoV has never been reported in horses in Israel. In recent years, other members of the group *β-coronavirus* have been detected in Israel in both animals and human. Antibodies against MERS-CoV were tested in sera samples that were collected from dromedary camels (*Camelus dromedaries*) between 2012 and 2017 and 61.8% had neutralizing antibodies against MERS-CoV (in virus neutralizing test, VNT) [18]. In samples that were collected from influenza-like illness patients in Israel in 2015–2016, HCoV-OC43, HCoV-NL63 and HCoV-229E were detected; however, no MERS-CoV infections were detected in human patients [19]. According to the Israeli veterinary services annual report in Hebrew, more recently, in 2019, 88/264 (33.3%) serum samples from camels were seropositive to MERS-CoV by ELISA, but 0/18 nasal swabs were positive by qPCR [20]. In 2019, 81/245 (33.1%) samples from cattle were also positive for BCoV by qPCR [20]. During the last year, like most countries in the world, Israel experienced a massive COVID-19 outbreak in which hundreds of thousands were infected and more than 5000 humans died. The aim of this study was to investigate the seroprevalence and selected risk factors for infection with ECoV among apparently healthy horses in Israel.

2. Materials and Methods

2.1. Study Population

Active surveillance and sera collection were conducted in 2018 and included 333 apparently healthy horses from 29 farms throughout Israel (4–32 at each farm) (Table S1). Horse owners approved the sample collection and the study protocol was approved by the Internal Research Committee of the Koret School of Veterinary Medicine–Veterinary Teaching Hospital (KSVM-VTH/08_2017). Fifteen farms and 150 horses were located in northern Israel, six farms with 93 horses were from central Israel and eight farms with 90 horses were from southern Israel. Almost half of the horses were mixed breeds (156, 46.8%) and others were of various breeds, including Quarter horses (65, 19.5%), Arabians (45, 13.5%), Ponies (19, 5.7%), Warmbloods (16, 4.8%) and Tennessee Walking horses (12, 3.6%). The study population included 161 mares (48.3%), 164 geldings (49.2%) and 8 stallions (2.4%). The horses' age ranged between six months and 47 years (mean = 11.66, median = 11, interquartile range (IQR) = 7). Most horses were kept in stalls (136, 40.8%), some were turned out in paddocks (133, 39.9%) and others were kept on pastures (19.2%, 64).

2.2. Sample Collection and Serologic Detection of ECoV Exposure

Blood samples were collected into a sterile tube without anticoagulant from the jugular vein. Serum was separated from each sample after centrifuging at 3000× *g* for 10 min and stored in −20 °C until analysis. Sera were shipped on ice and tested in the Department of Medicine and Epidemiology, School of Veterinary Medicine, University of California, Davis, CA, USA for the presence of antibodies against ECoV using the S1-based enzyme-linked immunosorbent assay (ELISA), as was previously described [16]. The S1-based ELISA targets antibodies to the spike (S) protein of ECoV and has been developed and validated using serum samples from naturally infected adult horses involved in contemporary outbreaks [16].

2.3. Statistical Analysis

Risk factors associated with exposure to ECoV (geographical area, horses' breed and sex and housing management) were assessed by using a chi-square test or two-sided Fisher's exact test, as appropriate, and the odds ratios were calculated. Association between animal age and exposure to ECoV was evaluated using a non-parametric Mann–Whitney U test. Factors that were found to be significantly associated with ECoV exposure in the univariable analysis were included in a multivariable analysis using the generalized estimating equation (GEE) with the farm set as a subject (i.e., random variable). The analysis was performed using SPSS 25.0® (IBM Corp, Armonk, NY, USA) and Win Pepi 11.65® statistical software (Abramson, J.H. WINPEPI updated: computer programs for epidemiologists, and their teaching potential. Epidemiologic Perspectives & Innovations, 2011, 8:1). Statistical significance was set as $p < 0.05$. A seroprevalence map was prepared using ArcGIS Desktop 10.6.19270 (ESRI, Redlands, CA, USA).

3. Results

3.1. Equine Coronavirus (ECoV) Seroprevalence

The seroprevalence of ECoV in the study population was 12.3% (95% confidence interval (CI): 8.98–16.33%), with 41 of the 333 horses testing positive for the presence of anti-ECoV antibodies. The seroprevalence in different farms varied between 0 and 37.5%. Exposure to ECoV was detected in 17 of 29 farms (58.6%). In most positive farms (11 of 17), exposure was identified in a single horse, while in six farms between two and 12 horses tested positive, with the positive farm prevalence ranging between 12.5% and 37.5% (Figure 1).

3.2. Risk Factors Associated with Exposure to ECoV

ECoV seroprevalence was higher in horses residing in central Israel than in horses from the north or south (odds ratio (OR) = 6.08, $p < 0.001$, Table 1), and was lower in horses kept in pastures (OR = 0.19, $p = 0.015$, Table 1). Although the median age was slightly higher in seropositive horses (13 years versus 11 years), the distribution of ages did not differ statistically between seropositive and seronegative horses ($p = 0.055$). A significant interaction was found between the geographical area and housing management, as the vast majority (63 of 64) of pastured horses resided in the North. The only factor found to be significantly associated with ECoV exposure in the multivariable model was horses residing in central Israel ($p < 0.001$).

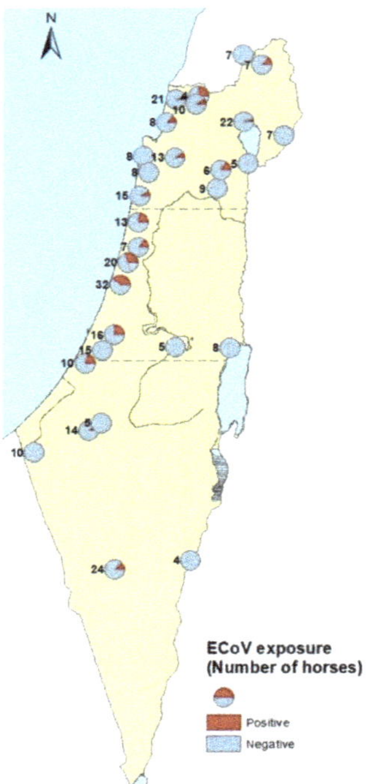

Figure 1. Geographical distribution of farms with horses tested for antibodies against Equine coronavirus (ECoV), indicating farms with only negative or ≥1 positive horse.

Table 1. Univariable analysis of the risk factors associated with exposure to Equine coronavirus (ECoV) for horses in Israel, 2018.

Variable	Category	*N*	ECoV-Positive (%)	OR (95% CI)	*p*
Area	North	150	9 (6%)	ref	-
-	Center	93	26 (28%)	6.08 (2.57–15.48)	<0.001
-	South	90	6 (6.7%)	1.12 (0.32–3.66)	1
Breed	Mixed	156	18 (11.5%)	ref	-
-	Pure bred	177	23 (13%)	1.15 (0.56–2.35)	0.74
Sex	Mare	161	21 (13%)	ref	-
-	Stallion	8	2 (25%)	2.22 (0.21–13.45)	0.298
-	Gelding	164	18 (11%)	0.82 (0.39–1.7)	0.611
Housing	Stall	136	20 (14.7%)	ref	-
-	Paddock	133	19 (14.3%)	0.97 (0.46–2.02)	1
-	Pasture	64	2 (3.1%)	0.19 (0.02–0.82)	0.015

4. Discussion

The seroprevalence for ECoV in horses in Israel was 12.3% (CI 8.98–16.33%) and varied between geographical locations, similarly to the report from the USA, where ECoV seroprevalence was 9.6%, and varied between 4.0% and 19.7% in different states [9].

In this study, the rate of exposure was significantly higher in horses residing in central Israel and significantly lower in horses housed on pastures. Considering that the majority of the horses housed on pastures were from the North of Israel, these two factors were associated, and it is not surprising that the seroprevalence was lower in the North. It is possible that the orofecal route of transmission is limited in horses on pastures, due to a lower density, open air, and natural ventilation. As previously described, increased density (i.e., small paddocks compared to pasture) is likely to increase orofecal transmission of ECoV [8,9,13,15].

While the majority of horses in central Israel live in stalls or paddocks, it is relevant to mention that some large farms experience a relatively high turnover of horses due to leisure and competition considerations. It is possible that the relatively higher density in this housing practice, together with the increased introduction of new horses (occasionally imported from the USA or EU), increase the risk of exposure to carrier individuals or environmental contamination [9,21].

The horse population in this study was comprised of apparently healthy horses with no reported recent history of illness. However, the clinical significance of ECoV infection in this population is unknown. In addition, the use of a serologic assay in a single blood test does not allow to estimate the time of exposure and cannot rule out past or future clinical signs. Moreover, since it is unknown how long after exposure the horses remain seropositive (assuming no repeated exposure), it is impossible to date the timing of exposure and associate it with the medical history of these horses. The fact that the majority of horses infected with ECoV remain subclinical or experienced only mild non-specific clinical signs, such as fever or enteric disease [22], also complicates any retrospective assessment of possible infection in "positive" farms.

Similar to other β-coronaviruses, ECoV is considered to originate from bats [22], possibly descending from BCoV or a rat coronavirus [22]. Thus, while the introduction of ECoV to Israel may have resulted from international horse trade and horses travelling between the USA, the EU and Israel, a possible local exposure to a bat coronavirus should also be considered. Regardless of their housing management, most horses in Israel are not isolated from the environment, either rural or urban (walls and windows are usually freely open to the environment); therefore, such exposure is highly plausible. Future characterization of the Israeli ECoV strains is needed to further investigate its possible origin and its enzootic potential or zoonotic risk.

The amino acid sequence of the ECoV Spike protein is considered to be highly conserved [7]. ECoV is closely related to the BCoV and camel coronavirus (HKU23) [23,24], and horses vaccinated with the BCoV vaccine were demonstrating some extent of neutralizing immunity against ECoV [24]. In the same year that the study serum samples were collected (2018), 20/52 (38.5%) of the fecal and intestinal samples from cattle that were tested for BCoV in the Kimron Veterinary Institute were positive by qPCR [25]. Since some horse farms in Israel are located near bovine stocks (dairy and beef pastures), exposure of horses to BCoV is possible and needs to be further investigated. On one of the farms in the south of Israel, where 3/24 horses tested seropositive for ECoV, horses were housed in paddocks within a large alpaca farm. A recent study indicated limited circulation of MERS-CoV in camelids in Israel [18]. Little is known of the possible circulation of coronaviruses between mammalian hosts and molecular studies are warranted to investigate possible common sources of infection in different animal species.

The ECoV S protein-based ELISA used in this study has a sensitivity of 100% and a specificity of 90.48% when using an OD cutoff of 1.958 [16]. Nucleocapsid (N) protein-based serological assays are more often associated with cross-reactivity than S protein-based assays [26], and yet, without molecular confirmation of ECoV in horses in Israel, cross-reactivity cannot be completely dismissed and the results of this study should be interpreted with caution.

This study indicates that horses in Israel were exposed to coronavirus, most probably to ECoV. Therefore, ECoV should be included in the differential diagnosis list of pathogens in cases of adult horses with anorexia, lethargy, fever and gastrointestinal signs in Israel. Continuous surveillance, the isolation and characterization of isolates, and the identification of the origin of infection is needed to further characterize the clinical and epidemiological significance of ECoV in horse populations.

5. Conclusions

Exposure of horses in Israel to coronavirus, most probably to ECoV, is low, with higher seroprevalence in horses residing in central Israel. So far, exposure of horses to ECoV in the Middle East was only reported from Saudi Arabia. Further surveillance and molecular characterization of ECoV in horses from Israel is needed to confirm its presence in the area.

Supplementary Materials: The following are available online at https://www.mdpi.com/2076-2615/11/3/894/s1, Table S1: Study population of horses from 29 farms throughout Israel.

Author Contributions: Conceptualization, A.S. (Amir Steinman), G.S. and N.P.; methodology, N.P. and S.B.; software, S.T.-L.; validation, S.B., A.S. (Asaf Sol), D.D. and N.P.; formal analysis, S.T.-L.; investigation, G.S., S.B. and S.T.-L.; resources: A.S. (Amir Steinman) and N.P.; data curation, G.S., S.T.-L. and A.S. (Amir Steinman); writing—original draft preparation, G.S. and S.T.-L.; writing—review and editing, G.S., S.T.-L., D.D., A.S. (Asaf Sol) and A.S. (Amir Steinman); supervision, A.S. (Amir Steinman) and N.P.; project administration, G.S. and S.B.; funding acquisition, A.S. (Amir Steinman) and N.P. All authors have read and agreed to the published version of the manuscript.

Funding: This research was funded by the Center for Companion Animal Health (CCAH), UC Davis and The Koret School of Veterinary Medicine (KSVM).

Institutional Review Board Statement: Sample collection was performed under the horse owners' consent, and the study was approved by the Internal Research Committee of the Koret School of Veterinary Medicine–Veterinary Teaching Hospital (KSVM-VTH/08_2017, 4 May 2017).

Data Availability Statement: Data is contained within the article or Supplementary Material.

Acknowledgments: The authors would like to thank Hadas Levi and Noa Masri for helping with sera collection.

Conflicts of Interest: The authors declare no conflict of interest.

References

1. Zhang, J.; Guy, J.S.; Snijder, E.J.; Denniston, D.A.; Timoney, P.J.; Balasuriya, U.B.R. Genomic characterization of equine coronavirus. *Virology* **2007**, *369*, 92–104. [CrossRef]
2. Krishnamoorthy, S.; Swain, B.; Verma, R.S.; Gunthe, S.S. SARS-CoV, MERS-CoV, and 2019-nCoV viruses: An overview of origin, evolution, and genetic variations. *VirusDisease* **2020**, *31*, 411–423. [CrossRef]
3. Pusterla, N.; Mapes, S.; Wademan, C.; White, A.; Ball, R.; Sapp, K.; Burns, P.; Ormond, C.; Butterworth, K.; Bartol, J.; et al. Emerging outbreaks associated with equine coronavirus in adult horses. *Vet. Microbiol.* **2013**, *162*, 228–231. [CrossRef]
4. Hemida, M.G.; Chu, D.K.W.; Perera, R.A.P.M.; Ko, R.L.W.; So, R.T.Y.; Ng, B.C.Y.; Chan, S.M.S.; Chu, S.; Alnaeem, A.A.; Alhammadi, M.A.; et al. Coronavirus infections in horses in Saudi Arabia and Oman. *Transbound. Emerg. Dis.* **2017**, *64*, 2093–2103. [CrossRef] [PubMed]
5. Wilson, J.H.; Cudd, T. Common gastrointestinal diseases. In *Equine Clinical Neonatology*; Koterba, A.M., Drummond, W.H., Kosch, P.C., Eds.; Lea & Febiger: Philadelphia, PA, USA, 1990; pp. 412–430.
6. Miszczak, F.; Tesson, V.; Kin, N.; Dina, J.; Balasuriya, U.B.R.; Pronost, S.; Vabret, A. First detection of equine coronavirus (ECoV) in Europe. *Vet. Microbiol.* **2014**, *171*, 206–209. [CrossRef]
7. Oue, Y.; Ishihara, R.; Edamatsu, H.; Morita, Y.; Yoshida, M.; Yoshima, M.; Hatama, S.; Murakami, K.; Kanno, T. Isolation of an equine coronavirus from adult horses with pyrogenic and enteric disease and its antigenic and genomic characterization in comparison with the NC99 strain. *Vet. Microbiol.* **2011**, *150*, 41–48. [CrossRef]
8. Giannitti, F.; Diab, S.; Mete, A.; Stanton, J.B.; Fielding, L.; Crossley, B.; Sverlow, K.; Fish, S.; Mapes, S.; Scott, L.; et al. Necrotizing Enteritis and Hyperammonemic Encephalopathy Associated With Equine Coronavirus Infection in Equids. *Vet. Pathol.* **2015**, *52*, 1148–1156. [CrossRef] [PubMed]
9. Kooijman, L.J.; James, K.; Mapes, S.M.; Theelen, M.J.P.; Pusterla, N. Seroprevalence and risk factors for infection with equine coronavirus in healthy horses in the USA. *Vet. J.* **2017**, *220*, 91–94. [CrossRef] [PubMed]
10. Oue, Y.; Morita, Y.; Kondo, T.; Nemoto, M. Epidemic of equine coronavirus at obihiro racecourse, Hokkaido, Japan in 2012. *J. Vet. Med. Sci.* **2013**, *75*, 1261–1265. [CrossRef] [PubMed]
11. Sanz, M.G.; Kwon, S.Y.; Pusterla, N.; Gold, J.R.; Bain, F.; Evermann, J. Evaluation of equine coronavirus fecal shedding among hospitalized horses. *J. Vet. Intern. Med.* **2019**, *33*, 918–922. [CrossRef] [PubMed]
12. Pusterla, N.; Vin, R.; Leutenegger, C.M.; Mittel, L.D.; Divers, T.J. Enteric coronavirus infection in adult horses. *Vet. J.* **2018**, *231*, 13–18. [CrossRef] [PubMed]
13. Pusterla, N.; Vin, R.; Leutenegger, C.; Mittel, L.D.; Divers, T.J. Equine coronavirus: An emerging enteric virus of adult horses. *Equine Vet. Educ.* **2016**, *28*, 216–223. [CrossRef] [PubMed]

14. Goodrich, E.L.; Mittel, L.D.; Glaser, A.; Ness, S.L.; Radcliffe, R.M.; Divers, T.J. Novel findings from a beta coronavirus outbreak on an American Miniature Horse breeding farm in upstate New York. *Equine Vet. Educ.* **2020**, *32*, 150–154. [CrossRef] [PubMed]
15. Fielding, C.L.; Higgins, J.K.; Higgins, J.C.; Mcintosh, S.; Scott, E.; Giannitti, F.; Mete, A.; Pusterla, N. Disease Associated with Equine Coronavirus Infection and High Case Fatality Rate. *J. Vet. Intern. Med.* **2015**, *29*, 307–310. [CrossRef]
16. Kooijman, L.J.; Mapes, S.M.; Pusterla, N. Development of an equine coronavirus–specific enzyme-linked immunosorbent assay to determine serologic responses in naturally infected horses. *J Vet. Diagn. Investig.* **2016**, *28*, 414–418. [CrossRef]
17. Zhao, S.; Smits, C.; Schuurman, N.; Barnum, S.; Pusterla, N.; van Kuppeveld, F.; Bosch, B.J.; Maanen, K.; Egberink, H. Development and validation of a S1 protein-based ELISA for the specific detection of antibodies against equine coronavirus. *Viruses* **2019**, *11*, 1109. [CrossRef]
18. David, D.; Rotenberg, D.; Khinich, E.; Erster, O.; Bardenstein, S.; van Straten, M.; Okba, N.M.A.; Raj, S.V.; Haagmans, B.L.; Miculitzki, M.; et al. Middle East respiratory syndrome coronavirus specific antibodies in naturally exposed Israeli llamas, alpacas and camels. *One Health* **2018**, *5*, 65–68. [CrossRef]
19. Friedman, N.; Alter, H.; Hindiyeh, M.; Mendelson, E.; Avni, Y.S.; Mandelboim, M. Human coronavirus infections in Israel: Epidemiology, clinical symptoms and summer seasonality of HCoV-HKU1. *Viruses* **2018**, *10*, 515. [CrossRef]
20. Balaish, M. Available online: https://www.moag.gov.il/vet/dochot-shnatiim/Documents/doch_shnati_2019.pdf (accessed on 1 February 2021). (In Hebrew)
21. Pusterla, N.; James, K.; Mapes, S.; Bain, F. Frequency of molecular detection of equine coronavirus in faeces and nasal secretions in 277 horses with acute onset of fever. *Vet. Rec.* **2019**, *184*, 385. [CrossRef]
22. Haake, C.; Cook, S.; Pusterla, N.; Murphy, B. Coronavirus Infections in Companion Animals: Virology, Epidemiology, Clinical and Pathologic Features. *Viruses* **2020**, *12*, 1023. [CrossRef]
23. Pusterla, N.; Vin, R.; Leutenegger, C.; Mittel, L.D.; Divers, T.J. *Equine Coronavirus Infection. Emerging and Re-emerging Infectious Diseases of Livestock*; Springer International Publishing: New York, NY, USA, 2017; pp. 121–132. [CrossRef]
24. Nemoto, M.; Kanno, T.; Bannai, H.; Tsujimura, K.; Yamanaka, T.; Kokado, H. Antibody response to equine coronavirus in horses inoculated with a bovine coronavirus vaccine. *J. Vet. Med. Sci.* **2017**, *79*, 1889–1891. [CrossRef] [PubMed]
25. Balaish, M. Available online: https://www.moag.gov.il/vet/dochot-shnatiim/Documents/doch_shnati_2016-2018.pdf (accessed on 1 February 2021). (In Hebrew)
26. Meyer, B.; Drosten, C.; Muller, M.A. Serological assays for emerging coronaviruses: Challenges and pitfalls. *Virus Res.* **2014**, *194*, 175–183. [CrossRef] [PubMed]

Review

Isothermal Nucleic Acid Amplification Technologies for the Detection of Equine Viral Pathogens

Alexandra Knox and Travis Beddoe *

Department of Animal, Plant and Soil Sciences and Centre for AgriBioscience La Trobe University, Bundoora, VIC 2082, Australia; a.knox@latrobe.edu.au
* Correspondence: T.Beddoe@latrobe.edu.au; Tel.: +61-3-9032-7400

Simple Summary: Equine viral diseases remain a prominent concern for human and equine health globally. Many of these viruses are of primary biosecurity concern to countries that import equines where these viruses are not present. In addition, several equine viruses are zoonotic, which can have a significant impact on human health. Current diagnostic techniques are both time consuming and laboratory-based. The ability to accurately detect diseases will lead to better management, treatment strategies, and health outcomes. This review outlines the current modern isothermal techniques for diagnostics, such as loop-mediated isothermal amplification and insulated isothermal polymerase chain reaction, and their application as point-of-care diagnostics for the equine industry.

Abstract: The global equine industry provides significant economic contributions worldwide, producing approximately USD $300 billion annually. However, with the continuous national and international movement and importation of horses, there is an ongoing threat of a viral outbreak causing large epidemics and subsequent significant economic losses. Additionally, horses serve as a host for several zoonotic diseases that could cause significant human health problems. The ability to rapidly diagnose equine viral diseases early could lead to better management, treatment, and biosecurity strategies. Current serological and molecular methods cannot be field-deployable and are not suitable for resource-poor laboratories due to the requirement of expensive equipment and trained personnel. Recently, isothermal nucleic acid amplification technologies, such as loop-mediated isothermal amplification (LAMP) and insulated isothermal polymerase chain reaction (iiPCR), have been developed to be utilized in-field, and provide rapid results within an hour. We will review current isothermal diagnostic techniques available to diagnose equine viruses of biosecurity and zoonotic concern and provide insight into their potential for in-field deployment

Keywords: equine; viruses; loop-mediated isothermal amplification; insulated isothermal polymerase chain reaction; field-deployable; point-of-care testing

Citation: Knox, A.; Beddoe, T. Isothermal Nucleic Acid Amplification Technologies for the Detection of Equine Viral Pathogens. *Animals* **2021**, *11*, 2150. https://doi.org/10.3390/ani11072150

Academic Editors: Amir Steinman and Oran Erster

Received: 30 June 2021
Accepted: 18 July 2021
Published: 20 July 2021

1. Introduction

Since their domestication, equines have been a pivotal part of history and continue to provide fundamental economic value worldwide [1–3]. With an estimated global population of over 59 million domesticated horses [4], the global equine industry is valued at approximately USD $300 billion annually [2,5]. The industry comprises of two main categories: primary equine activities and secondary equine activities. Primary activities are defined as sectors directly involved with equines, such as horse trainers, coaches, breeders, professional competitors and jockeys, and clubs and associations. In contrast, the secondary sector is for services that are indirectly involved with equines, or provide external services for equine owners, such as equine health professionals, and support industries including transport and sale of horses [6]. These sectors provide essential services for countries worldwide, significantly contributing to strong economic growth, particularly in developing communities [3,7].

In addition to the economic contributes, the global equine industry has an estimated 1.6 million full-time employees. In particular, the racing industry is the major contributor with significant levels of employment, from trainers and jockeys to breeders [8]. With over 160,000 races held worldwide annually [9], the economic substance of this industry is apparent. Additionally, the racing industry provides longstanding culture and traditions throughout the world. For example, the Melbourne Cup, held in Australia, is the most renowned handicap Thoroughbred equine racing event of the year [10]. Over 22 countries participate and import their Thoroughbreds to Australia for the racing seasons, reaching a yearly global audience of over 700 million [11].

While the equine industry is extremely important economically and socially, as either organized equine sport or companion animals, there is a range of zoonotic and non-zoonotic viral infections that are harmful to both equine and human health [12–14]. For example, Australia experienced an outbreak of equine influenza in 2007, affecting roughly 69,000 horses and resulting in a significant economic loss estimated at a current AUD $571 million, with eradication alone costing an inflated $370 million [15]. Fortunately, Australia was able to eradicate this virus; however, further worldwide viral outbreaks continuously loom over the fate of the industry [14]. With continuous global movement, importation and subsequent housing of large equine populations increasing worldwide, it is essential to increase biosecurity measures and diagnostics against viral diseases to avoid rapid transmission and spread [16].

Moreover, many of these diseases do not have effective treatment options; thus, there is an increased demand to control and eradicate diseases through improved biosecurity protocols [12,17,18]. The ability to accurately diagnose diseases early could lead to better management and treatment strategies [16]. Diagnostic methods have been developed over previous decades due to advances in biochemistry, molecular biology, and immunology research [19] and continue to improve presently. These advancements, such as and point-of-care (POC) diagnostics, are increasingly utilized and sought after for routine diagnosis for equine viral infections [16]. While many molecular tests, such as polymerase chain reaction (PCR), have been developed to detect equine viral infections, they are not field-deployable, thus are unable to support rapid decision-making for disease control and treatment [20–22]. To overcome these drawbacks current research has moved toward isothermal nucleic amplification techniques, such as loop-mediated isothermal amplification (LAMP) [23] and insulated isothermal polymerase chain reaction (iiPCR) [24]. Both these methods utilize an enzymatic reaction to amplify nucleic acid, at a constant temperature [23,24]. LAMP and iiPCR have been previously demonstrated to be field deployable POC diagnostic techniques, achieving results in less than an hour. These powerful tools have been extensively researched for equine medicine, and continue to pave the way for newer, more accessible diagnostic methods [23–29]. Here we review the field-deployable technology, LAMP and iiPCR, and their application to diagnose equine viral infections.

2. Equine Viral Diseases of Biosecurity Concern

Despite strict global import and biosecurity policies, infectious disease outbreaks continue to occur globally, particularly with equine viruses [13,30]. These outbreaks have detrimental effects on the equine's health and welfare and inhibit their regular activity, subsequently harming the industry's economy in the associated geographical regions [15,17,20]. The World Organisation of Animal Health (OIE) releases a yearly report stating the diseases of concern for terrestrial animals, which includes equine viral pathogens [31,32]. This section outlines each of these viral diseases.

2.1. African Horse Sickness

African horse sickness (AHS) is a non-contagious arthropod-borne virus widely distributed across sub-Saharan Africa [33]. There are four forms of the disease: subclinical, subacute or cardiac, acute respiratory, and mixed. Mortality rates vary with disease severity, with the mixed and acute respiratory forms having the highest mortality rates at 70–80%

and 95%, respectively [34]. As AHS is transmitted to a susceptible host via a mosquito vector, mainly *Culicoides* species, the virus can quickly spread before containment [35,36]. Moreover, recent studies have warned that the distribution of AHS is expanding from endemic areas to regions with suitable climatic environments that are home to other mosquito species which share ancestry with *Culicoides* species [35–37]. In fact, four horses in Thailand during March 2020 tested positive for AHS after succumbing to infection just 12–24 h after initially displaying symptoms, making quick diagnosis paramount [38]. Furthermore, the government had to quickly implement control measures and utilize live attenuated vaccines [39]. Despite the availability and the continuous development of AHS vaccines [40], many countries including Australia, still do not have approval for implementation to these options [41], leaving them vulnerable to a potential outbreak without a means to control the disease [36].

2.2. Equine Encephalomyelitis (Western)

While western equine encephalitis (WEE) persistence has been declining considerably since the mid-20th century [42–45], this arbovirus remains on the OIE list of notifiable diseases [31,32]. The choice to continually survey for this virus is attributed to the potential for further significant and detrimental outbreaks [42]. The virus circulates in an enzootic cycle between mosquitoes, specifically *Culex* species, and passerine birds. However, infection of humans and equines can occur in the event of a spillover during peak vector activity periods [42,46–48]. Cases have declined since the 1940s and 1950s, which saw peak cases in humans and equines in America's western region [42]. Clinical signs in horses start with biphasic fever, followed by a range of neurological and behavioural symptoms, including anorexia, ataxia, aggression, somnolence, aimless wandering, general depression, and animals eventually succumb to the disease [49–51]. In humans, WEE infections can result in neurological sequelae post-infection which places a severe strain on health care system. Treatment costs for human infection varies between $21,000 and $3 million per case [42,52]. There is no specific antiviral treatment for both humans and equines, with supportive care the only available option [48,51,53]. The recommended diagnostic techniques for WEE include virus isolation and reverse-transcription PCR (RT-PCR) [31]; however, development is in progress for a nucleic acid sequence-based amplification (NASBA) assay that could provide a more rapid means of detection and be used for field samples [54]. However, this assay is yet to be validated.

2.3. Equine Infectious Anaemia

Equine infectious anaemia (EIA) is a non-contagious disease of equids; however, equines and ponies are more susceptible to severe clinical infection of this virus. This globally prevalent disease causes all infected equids to become life-long carriers [55–58]. Transmission occurs through blood-feeding vectors, specifically horseflies and deerflies, blood-contaminated fomites, and in utero via transplacental transmission [55,59,60]. Clinical signs vary depending on the strain virulence and susceptibility of the equid host. Majority of cases occur in three phases; the acute or initial phase, followed by a chronic phase, and finally the inapparent, or long-term asymptomatic phase [55,57,61]. Clinical symptoms typically appear within seven to thirty days post-infection, with fever, depression, and possible thrombocytopenia; however, signs may be mild and can be overlooked, resulting in misdiagnosis or underreporting [55,57,62]. Equines experience reoccurring episodes of fever, increased heart and respiratory rates, anaemia, muscle weakness, and loss of condition for around one year following initial infection [55]. Equines will then become chronic life-long carriers with no apparent symptoms [57]. EIA has caused severe outbreaks throughout Europe and has re-emerged in countries after multiple years of disease absence [63]. Diagnostics are exclusively performed by serological techniques, including enzyme-linked immunosorbent assay (ELISA) [59,61,64]; however, only the agar gel immunodiffusion (AGID) assay remains OIE approved [31,59]. Despite this recommendation, the AGID assay can require a secondary test for validation [31] and is not appropriate for equines in the acute phase of infection as viral load is too low [65].

2.4. Equine Influenza

Equine influenza (EI) is reported as the most important globally distributed respiratory disease in equines [66]. The fatality rates are contributed to the secondary bacterial infection; however, the prognosis typically relies on the individual immune status [67]. EI is highly contagious and has multiple transmission pathways, including contaminated fomites. Furthermore, there is no specific treatment, and despite an available vaccine, significant outbreaks continue to occur [66]. As previously stated, Australia experienced an EI outbreak in 2007 that lasted for five months affecting roughly 69,000 horses [15,68]. The magnitude of the outbreak affected 9,600 properties, including companion equine households, business incomes, and horse associations [69]. The strict biosecurity measures were implemented and remain ongoing; however, a recent survey of 1,224 horse owners directly involved in the 2007 outbreak reported that 32% of participants were not in favor of continuous biosecurity measures. More concerningly, approximately 30% of participants had low biosecurity compliance, stating they implemented biosecurity procedures "not often" or "never" [30]. This complacency, or lack of understanding, further enhances the risk of outbreaks throughout the equine industry [70]. More recently, the United States has had waves of annual epidemics in 2015, 2016, and 2017, affecting 23, 16, and 22 states, respectively [66].

Additionally, in 2018, Chile experienced a re-emergence of the H3N8 EI strain, which had not previously been detected since 2012. Further genetic testing confirmed that this virus had high homology with other viruses that had been in circulation in Europe and Asia [71]. It is apparent that EI is continuously present almost globally, and outbreaks will continue to fluctuate without adequate means of rapid diagnostics to quickly and efficiently intervene [66,70,72].

2.5. Equine Viral Arteritis

Equine viral arteritis (EVA) significantly impacts the breeding sector of the equine industry, as the disease affects both the respiratory and reproductive status of the animal. EVA incidences have been increasing over the past 20 years [73]. While the majority of cases are subclinical, serious long-term effects cause significant production losses [74]. The disease is rarely fatal in healthy horses; however, 50%–60% of infected pregnant mares can experience abortion [75,76]. In addition, stallions can be long-term carriers while remaining asymptomatic [74]. Like many other equine viral diseases, there is no affective treatment, restricting countries to rely on biosecurity measures [77]. EVA can be spread through venereal mechanisms; leaving breeding programs at a high transmission risk, which is a prominent industry in many countries. Long-term carrier stallions may have to undergo castration to prohibit accidental transmission to mares, and removing them from breeding programs [77,78]. The disease has already caused a reduction in commercial value of horses, with higher costs for breeding and commercialization of semen and embryos [77]. This was particularly evident in 2007 when France experienced an outbreak of EVA due to the distribution of infected semen, causing the disease to spread into 17 premises. It was suspected that horizontal transmission occurred via farm employees. This outbreak was deemed the most significant of its kind, with considerable economic disruptions [79]. While the use and advocation of vaccine programs can help alleviate the burden, persistently importing infected equines remains highly problematic for vulnerable countries [73].

2.6. Equine Rhinopneumonitis (Caused by EHV-1)

Equine rhinopneumonitis caused by equine herpesvirus type 1 (EHV-1) is globally distributed, particularly in regions with a significant equine presence [80–82]. Additionally, equine rhinopneumonitis can be caused by equine herpesvirus type 4 (EHV-4), furthering exacerbating the prevalence of this disease [80,83,84]. Despite the availability of both live and inactivated vaccines for EHV-1, the persistence of this virus remains [85,86]. While transmission is predominantly via the respiratory route, contact or ingestion of contaminated fomites or contact through foetuses or placenta of an infected mare is possible [81,82,85,87]. Due to the inapparent respiratory clinical signs, it is often misdiagnosed as other viral or bacterial diseases, leaving equine populations susceptible to the introduction of the virus [82,88].

Additionally, younger equines appear to be highly susceptible to infection, with 80%–90% of animals less than two years old carrying this respiratory disease [89]. While non-steroidal anti-inflammatory drugs (NSAIDs) may assist in elevating symptoms, there is no specific cure for disease elimination [85,89]. Despite the sporadic recovery from infection, horses can often develop a secondary infection that can be fatal [90]. Despite the development of a PCR diagnostic assay, virus isolation is still required for comparative analysis to other diseases, making accurate disease identification laborious [31,84].

3. Zoonotic Equine Viral Diseases of Concern

3.1. Hendra Virus

Hendra virus (HeV) is a well-documented zoonotic equine virus that has been a prominent concern for the equine industry [91]. While this emerging, highly transmissible virus is exclusively isolated to Australia, it has caused several outbreaks and is predominantly fatal. The primary vector of HeV is fruit bats; although the exact mechanism of transmission to equines is not fully understood. However, it is thought that equines potentially consume contaminated fruit bat droppings via their feed. Transmission among equines and subsequently to humans is through either direct (via secretions) or indirect (via fomites) routes [92]. The disease presents as influenza-like symptoms with rapid deterioration [93]. There is no specific treatment or cure for HeV, and progression can lead to septic pneumonia, and more recently found, encephalitis [91,94]. In 2008, five equines and two human infection cases occurred in Queensland, Australia. As a result, many veterinary clinics had to close due to the ramifications of acquiring the virus [95]. Despite the currently available vaccine targeting equine HeV, there is no vaccine available for humans, leaving all equine industry personnel vulnerable to infection. The suggested prevention for human infections to avoid infected horses and maintaining personal hygiene [91,93,96]. With the limited feasibility to this approach it is reasonable to expect another HeV outbreak. An outbreak of HeV would infer severe economic losses from the cancellation of events and prohibition of animal movement [91,96].

3.2. Japanese Encephalitis

Typically known as a significant human neurological disease, the Japanese encephalitis virus (JeV) also infects equines with three clinical syndromes: transient, lethargic, and hyperexcitable type [97,98]. Horses infected with either transient or lethargic type typically recover within a week; however, death is common with the hyperexcitable type [99]. In addition to encephalitis, clinical signs in equines can also include a fluctuating fever, decreased appetite, jaundice and haemorrhaging in the mucous membranes, staggering, and sweating [97]. While this disease is not globally distributed, many populated countries in Asia encounter a combined 70,000 human cases per year with 10,000 of these being fatal [100]. Limited barriers separating endemic and JeV-free countries, coupled with the ease of mosquito vector transmission and limited availability of vaccines in non-endemic countries, make the risk of outbreaks significantly high [101]. Additionally, accurate detection of viral prevalence is problematic due to a short duration of viraemia and asymptomatic infections [100].

3.3. Ross River Virus

Ross River virus (RRV) is the most widespread and significant arbovirus in Australia and neighboring islands, such as Fiji and the Cook Islands, frequently causing large epidemics in humans and equines [102]. There has been an increase in incidences of infection across Australia due to recent flooding and climatic changes optimal to harbor the mosquito vector [103–105]. RRV can infect equines and humans through mosquito bites and causes various symptoms ranging from distal limb oedema and arthritis to neurological diseases [106–108]. Additionally, infected equines reluctantly move during infection due to debilitating joint pain, causing a significant reduction in production and performance [108]. The prescribed treatment for equines includes NSAIDs therapy considering there is no available vaccine [109]. The majority of infection reports state that recovery on average takes two to five days; however, recently prolonged recoveries of up to five months to a year have been noted. Not only is RRV a significant concern for human and equine health, the equine industry could also infer potential economic losses in the millions, attributed to restrictions on movement and trade, loss of performance in infected equines, and wastage [107]. Australia's favorable environmental and ecological conditions have facilitated an endemic state that encounters reoccurring outbreaks; with a likelihood of climatic change enhancing the global occurrence of optimal conditions for the spread of disease. Subsequently, increased outbreaks would cause the implementation of strict biosecurity measures to ensure both human and animal welfare; resulting in restrictions on animal movement, production, and quarantining in turn causing undoubtable economic losses [106].

3.4. West Nile Virus

West Nile virus (WNV) is closely related to JeV; however, seldom causes encephalitis in humans and equines [110,111]. In fact, most infected humans will be asymptomatic, with only around 20% of cases resulting in influenza-like symptoms [112,113]; nevertheless, viral infections can still be fatal [114]. For equines, clinical signs can include neurological disease, such as encephalitis and ataxia, as along with the loss of appetite, depression, and, infrequently, fever [115–117]. While there is a vaccine available for equines [118,119], the risk for expansive transmission and cross-species spread is foreseeable, due to the broad range of hosts, such as reptiles, mammals, birds, and ticks [120]. The main transmission mechanism is via carrier mosquitoes after biting an infected host, namely the Corvidae family of birds [112]. A mosquito then can infect several animals and bird species, including equines and humans, which are incidental hosts [120]. Jointly with the ease of transmissibility, WNV has a wide geographical distribution throughout Africa, Europe, West Asia, Australia, and North America, giving a high probability of global spread [110,121]. Considering there is no specific WNV treatment, supportive care is recommended until the infection subdues, typically spontaneously [111]. Currently, detection relies on nested and real-time reverse transcription PCR (real-time RT-PCR) [31,122,123]. However serological diagnosis, such as seroconversion, is more reliable, as current molecular tools are unable to provide accurate diagnostics due to their sensitivities and the low viremia associated with WNV infections [31]. While control programs are dependent on surveillance, particularly of deceased crows [120,124] and vaccine for equines [118,119], this does not entirely protect humans [120]. Ultimately, the concurrent broad host range and vast geographical distribution of WNV has the potential for a global outbreak with significant impact [121,125].

4. Current Diagnostic Techniques for Equine Viral Diseases

Diagnostics in the equine industry are vital to restrict the spread of infectious diseases [19], particularly with frequent and high equine movement [16]. Due to this substantial amount of transport nationally and internationally, the OIE has provided a list of 117 diseases of concern for terrestrial animals; six of these include equine viruses (Table 1) [32]. Additionally, OIE has produced a reference guide for terrestrial animal diagnostics to promote the use of "gold-standard" testing worldwide [31]. Table 1 presents the current gold standard diagnostic techniques for each of the named equine viruses and zoonotic equine viruses of concern.

Table 1. OIE [32] notifiable equine viruses and zoonotic viruses of biosecurity concern with their prescribed "gold-standard" diagnostic tests for confirmation of disease [31].

Disease	Prescribed Diagnostic Test/s [31]
OIE listed notifiable equine viral diseases [32]	
African horse sickness	RT-PCR [1] Virus isolation
Equine encephalomyelitis (Western)	RT-PCR Virus isolation
Equine infectious anaemia	AGID [2]
Equine influenza	ELISA [3] RT-PCR
Equine viral arteritis	CF [4] PCR VN [5] Virus isolation
Equine viral rhinopneumonitis (EHV-1)	PCR VN Virus isolation
Zoonotic equine viral diseases of concern	
Hendra virus	RT-PCR Virus isolation
Japanese encephalitis	RT-PCR Virus isolation
Ross River virus	RT-PCR [126] Virus isolation [126]
West Nile virus	RT-PCR

[1] Reverse-transcription polymerase chain reaction, [2] agar gel immunodiffusion assay, [3] enzyme-linked immunosorbent assay, [4] complement fixation, [5] virus neutralization.

4.1. Serological Diagnostics

Serological assays are used for an array of diagnostics in equine medicine, including viral diseases. These assays detect antibodies of a specific infection from the serum, providing an indirect mean of diagnosis [127]. Serological analysis is commonly utilized in equine medicine for diagnostics, due to the attractive advantages they provide mboxciteB16-animals-1300066,B31-animals-1300066,B127-animals-1300066. The use of serological assays allows for detection of samples with low quantity of antigens, visualization with the naked eye, and can also be used on a wide range of pathogens [16,127]. However, drawbacks of these assays must be considered. Firstly, many of these assays have a high probability of either false-positive or false-negative results [127]. Secondly, serological assays often are required be coupled with a secondary detection method for an official confirmation. Additionally, as with other types of diagnostics, the assays often require specialized equipment, and are time consuming and labor intensive, either due to the assay procedure or subsequently from a secondary diagnostic test for confirmation [16,128,129]. However, despite these drawbacks these assay techniques remain a well-established technique in veterinary medicine [127]. One common diagnostic technique is ELISA, an assay that detects specific immune responses with the use of antibodies, antigens, and enzymes [16]. ELISA is considered a convenient, safe, and reproducible diagnostic technique, with several different variations, such as dot-ELISA and falcon assay screening test-ELISA (FAST-ELISA). These developments have allowed for a quicker assay that is cost-efficient with results that can be easily interpretative [129]. ELISA has been proven as a reliable diagnostic tool for equine influenza, and its use is advocated by

OIE [31,127]. Despite serological assays being well-established in veterinary diagnostics, many of these assays are being replaced with newer molecular technologies [16].

4.2. Molecular Diagnostics

Molecular diagnostics have been continuously evolving, providing more sensitive detection of nucleic acid [130]. As a result, these tools have been increasingly favored and utilized in equine medicine for clinical diagnostics. PCR has been the most advocated molecular tool, with the greatest success [16]. PCR tests can detect various organisms, including slow-growing or challenging to cultivate organisms, overcoming limitations in previously used diagnostic tools [131]. Furthermore, PCR provides other advantages such as rapid time to gain results, the sensitivity to detect smaller quantities of microorganisms and is not reliant on the host's immune response [54,132,133].

Nevertheless, this method comes with drawbacks, including reaction inhibition from substances within samples, such as urea, varying techniques and protocols, frequency of false-negatives and false-positives, a high risk of contamination, and the requirement of expensive equipment and experienced personnel [54,134,135]. Additionally, as PCR is based on nucleic acid amplification, the results can only confirm the presence or absence of pathogenic DNA in the sample [136,137]. Yet, PCR is still considered a powerful tool that is utilized consistently in equine medicine [138]. Advancements in PCR-based technologies have been developed throughout recent years and have expanded diagnostic capabilities for detecting clinical infections, particularly for viruses [139].

Recently, PCR has been developed for real-time evaluation of results by utilizing intercalating dyes or target-specific probes [140], minimising handling of PCR products throughout the procedure, therefore reducing the risk of contamination [141]. In addition, many PCR assays have been described to utilize real-time PCR coupled with reverse-transcription, termed real-time RT-PCR [138]. Thus, this technique is now quickly replacing diagnostics that were previously [31,142,143] performed by conventional PCR.

Molecular diagnostics are a promising tool for detection of viral infections, but they can be misinterpreted by inexperienced personnel and are not applicable for in-field use or in poorly resourced laboratories, limiting their global disease surveillance application [138].

5. Isothermal Techniques

Isothermal techniques are driven by enzymatic reactions to amplify nucleic acid at a single temperature, thus allowing POC or field-deployable testing [144,145]. Additionally, some of these techniques do not require samples to be purified, allowing for direct use of living cells from field obtainable samples [145]. This advantage has influenced diagnostic technique development to further exploit isothermal conditions over conventional methods such as PCR, which requires various temperature cycles to complete amplification. Multiple isothermal technologies are currently available, with unique features and template types (Table 2) [146].

Despite promising and extensive research into isothermal techniques for pathogenic detection, equine diagnostic technology for viruses has been limited to two isothermal technologies, LAMP and iiPCR (Figure 1). This is probably due to several different companies producing commercial reagents for both LAMP and iiPCR assays. This has allowed researchers to develop assays for the detection of different viruses. However, the benefits of lower costs, low energy requirement, method simplicity, and ease of field deployment of isothermal technologies justify further research for diagnostics for equine medicine [23,24,149,155,156].

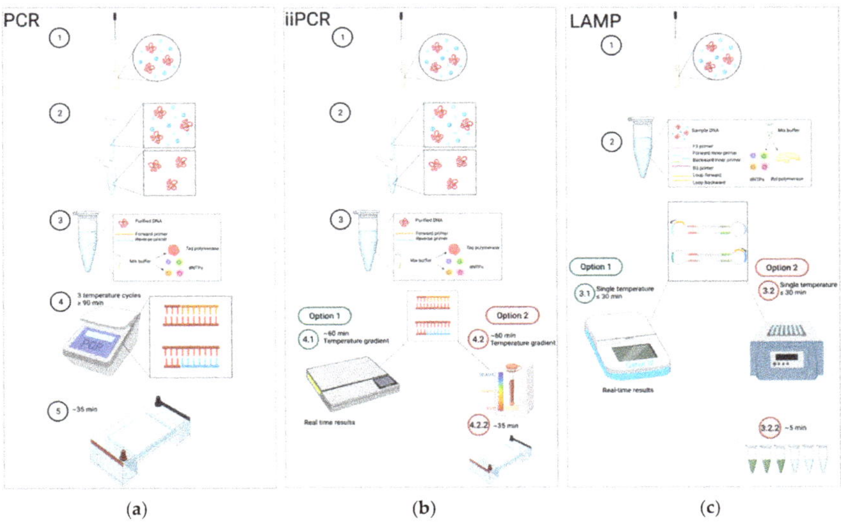

Figure 1. Comparison of polymerase chain reaction (PCR) insulated isothermal polymerase chain reaction (iiPCR) and loop-mediates isothermal amplification (LAMP) procedures. (**a**) PCR procedure is as follows; 1. sample is collected; 2. sample is purified; 3. contents for PCR are mixed including the purified sample, forward and reverse primers, and master mix buffer which includes *Taq* polymerase and dNTPs; 4. the reaction is ran on a thermocycler for ≥90 min cycling through three temperatures for the denaturation, annealing and extension stages; 5. PCR products are subjected to agarose gel electrophoresis for approximately 35 min at 100 amps to visualize results. (**b**) iiPCR follows a similar starting procedure to PCR where, 1. samples are collected, and 2. purified, 3. contents are mixed as such for PCR. However, reaction is conduced within capillary tubes with a copper ring at the base and lid, where the mixture is heated underneath to create a temperature gradient through convection; reactions last for around 1 h. This can be achieved through two options: 4.1. an automated portable machine, POCKIT™ (GeneReach USA, Lexington, MA, USA) where results are displayed in real time; alternatively, 4.2. an insulated box that requires the products to undergo (4.2.2) agarose gel electrophoresis for approximately 35 min at 100 amps to visualize results. (**c**) The LAMP procedure is as follows, 1. samples are collected and 2. mixed with 4–6 primers (F3, B3, forward inner primer and backward inner primer, and optional loop primers). LAMP can tolerate impurities in samples and therefore do not required to be purified. 3. The mixture is heated at a single temp temperature for typically ≤30 min. This can also be achieved by two options: 3.1. an automated machine, Genie III™, (OptiGene Horsham, Eng, UK), where results are displayed in real time; alternatively, 3.2. a heat source, such as a water bath, where products are visualized through (3.2.2) fluorescence for approximately 5 min to observe a color change. Created with BioRender.com.

Table 2. Summary of developed isothermal techniques.

Technique	Template	Temperature [1]	Enzyme	Reference
Helicase-dependent amplification (HDA)	DNA	65 °C	Helicase	[147]
Insulated isothermal PCR (iiPCR)	DNA	95 °C	Taq DNA polymerase	[24]
Insulated isothermal reverse-transcription PCR (iiRT-PCR)	RNA	95 °C	Taq DNA polymerase M-MLV reverse transcription	[148]
Loop-mediated isothermal amplification (LAMP)	DNA	65 °C	*Bst* DNA polymerase	[23]

Table 2. *Cont.*

Technique	Template	Temperature [1]	Enzyme	Reference
Reverse transcription loop-mediated isothermal amplification (RT-LAMP)	RNA	65 °C	*Bst* DNA polymerase AMV reverse transcription	[149]
Multiple displacement amplification (MDA)	DNA	30 °C	Φ29 DNA polymerase	[150]
Nucleic acid sequence-based amplification (NASBA)	RNA	50 °C	T7 RNA polymerase RNase H AMV reverse transcription	[151]
Rolling circular amplification (RCA)	DNA	30 °C	Phi29 *Bst* DNA polymerase Vent *exo*-DNA polymerase T7 RNA polymerase	[152]
Recombinase polymerase amplification (RPA)	DNA RNA	37 °C	DNA polymerase	[153]
Strand displacement amplification (SDA)	DNA	60 °C	DNA polymerase	[154]

[1] Average temperature used in respective assays.

6. Application of LAMP for Equine Viral Diseases

6.1. Principles of LAMP

LAMP was designed to overcome associated drawbacks of traditional serological and molecular diagnostics. Unlike other assays, LAMP does not require expensive equipment, trained personnel, laborious methods making it easily deployed in resource-poor settings [23]. This technique is relevant for various applications, such as rapid, sensitive, and specific diagnostics, genetic, and POC testing [20,21]. In addition, the DNA template does not need to be denatured, which is a requirement of conventional PCR [156], and results can be visualized with the naked eye. This reduces the number of required steps and subsequent downstream processing time and the possibility of cross-contamination, an issue common to other diagnostic techniques [157,158]. LAMP utilizes four to six primers that recognize six to eight distinct regions of a target sequence, enhancing the rapidity of the assay which is performed at a constant temperature (Figure 1) [23]. This application has proven to be a reliable diagnostic technique for a diverse range of pathogens, including equine infectious diseases [21,25,26,159,160].

6.2. Application of LAMP for Equine Viral Diseases

Due to the wide success of the currently available LAMP assays, there is continuous development of this technology for various applications. One such technique is incorporating a reverse-transcription to detect RNA viruses, coined RT-LAMP [27,149,161,162]. In addition to conventional LAMP, this approach has been utilized for numerous equine viral disease (Table 3).

Table 3. Current LAMP assays developed for equine viral diseases.

Disease	Type	Vector-Borne	Target Gene	Sample	Detection Limit	In-Field	Ref
African horse sickness	dsRNA	Yes—Midges, Mosquito	Vp7	Horse—Blood	n/a	Yes	[163]
Equine herpesvirus 1	dsDNA	No	Glycoprotein C Glycoprotein E	Horse—Nasal swab [1] Horse—Nasal swab [1]	1 pfu/rxn 1 pfu/rxn	No [1]	[164]

Table 3. *Cont.*

Disease	Type	Vector-Borne	Target Gene	Sample	Detection Limit	In-Field	Ref
Equine herpesvirus 4	dsDNA	No	Glycoprotein C	Horse—Nasal swab [1]	1 pfu/rxn	No [1]	[164]
Equine infectious anaemia	ssRNA	Yes—Horse and deer flies	Gag nsP	Recombinant plasmid	0.1 pfu/rxn	No	[165]
Equine influenza (H3N8)	ssRNA	No	HA	Horse—Nasal swab	10^{-5} copies/rxn	Yes	[166]
Equine influenza (H7N7)	ssRNA	No	HA	Horse—Nasal swab	10^{-4} copies/rxn	Yes	[167]
Equine coronavirus	ssRNA	No	Nucleocapsid	Horse—Nasal swab, fecal samples	$10^{1.8}$ copies/rxn	Yes	[168]
Hendra virus	ssRNA	No	P	Horse—Nasal swab [1]	10^{-5} copies/rxn	No [1]	[26]
St Louis encephalitis	ssRNA	Yes—mosquito	UTR	Mosquito	<0.1 pfu/rxn	Yes	[159]
Western equine encephalitis	ssRNA	Yes—mosquito	nsP4	Mosquito	100 pfu/ml	Yes	[159]
West Nile virus	ssRNA	Yes—mosquito	E	Mosquito	0.1 pfu/ml	Yes [2]	[149]

[1] Experimentally infected animals, [2] secondary experiment.

Nemoto et al. [164] developed a LAMP assay to detect both equine herpesvirus type 1 (EHV-1) and 4 (EHV-4), as well as differentiating between the wild-type EHV-1 (ΔgE) strain, which is the non-neuropathogenic strain [169]. This assay detected glycoprotein C (gC) in both viruses for diagnostic purposes and EHV-1 glycoprotein E (gE) for distinction from the wild-type strain, which has a deletion at the gE gene. This assay reported similar sensitivity compared to PCR, but at a lower cost, and a time to positive between 30 min to 1 h when ran at a constant temperature of 60–65 °C. The results were visualized by gel electrophoresis and by eye through observation of a color change. The detection limit for EHV-1 and EHV-4 showed high sensitivity at 1 and 0.1 plaque-forming unit (pfu) per reaction, respectively, with no cross-reaction towards other viral and bacterial equine diseases. Therefore, this LAMP assay has the potential to replace current PCR diagnostic assays to accurately determine equine herpesvirus [164].

RT-LAMP was developed to detect RNA viruses, which performs synthesis of DNA for detection concurrently with amplification [149]. This technique was also adopted by Nemoto and colleagues to develop two novel assays detecting equine influenza strains H3N8 [166] and H7N7 [167]. Both assays were designed to target the HA gene of influenza from nasal swab samples acquired in the field from horses presenting with a fever ($\geq$38 °C). The assays were specific to differentiate the separate strains. The assay was 3 to 10 times more sensitive than the commercial serological ELISA test (Espline Influenza A&B-N ELISA test (Fujirebio, Japan)).

Additionally, H3N8 RT-LAMP assay was ten times more sensitive than the previously developed RT-PCR test, while the H7N7 RT-LAMP compared the same as the RT-PCR. The H3N8 assay detected 35 additional positive samples that were not positively identified by both the RT-PCR and the Espline Influenza A&B-N test. The detection limit for H3N8 and H7N7 during RT-LAMP was 10^{-5} and 10^{-4} copies per reaction, respectively, achieved a positive threshold in roughly 60 min. The results of these assays were visually determined by turbidity, allowing for identification without specialized equipment. This approach for detection shows the simplicity that LAMP assays offer and the ability for in-field diagnostics and large-scale surveillance. The authors recommend combining these RT-LAMP assays into a panel diagnostic test to differentiate between the two strains [166,167].

Furthermore, Fowler et al. [163] utilized RT-LAMP to develop rapid detection of African horse sickness (AHS), resulting in a similar sensitivity to a previously developed AHS real-time RT-PCR. By targeting the structurally conserved VP7 gene that forms the outer capsid, the assay could detect the viral DNA within 30 min. The results were visualized by DNA intercalating dyes, contained in the reaction master mix (ISO-001, OptiGene Ltd., Horsham, UK). Despite the ease of visualization, the paper suggested adapting the assay to use real-time fluorescence for ease of application in-field, adopted from previous experiments [170,171]. To convert the assay to an in-field diagnostic technique, it was recommended to use lyophilized reagents and eliminate the RNA extraction procedure by implementing an automated extraction procedure, as utilized by Waters et al. [170] and Howson et al. [171].

In 2018, Han et al. [165] presented a preliminary study of a RT-LAMP assay for equine infectious anemia. This study employed detected the gag non-structural protein (gag nsP) of the virus, using a recombinant plasmid, pMD-19T-gag, rather than field or clinical samples. While this assay has a longer reaction time of two hours to detection 100 copies/μL, it provides a starting point for further development. As this assay only included four primers, it is possible to decrease the assay time through the use of loop primers. Promisingly, the RT-LAMP assay did not detect other pathogens, showing high specificity, which can be further validated through the testing of clinical samples. Furthermore, results were visualized through a color change, allowing for the possibility of conversion to a field deployable diagnostic technique.

Wheeler et al. [159] developed a panel of RT-LAMP assays for the detection of St. Louis encephalitis virus (SLEV) and western equine encephalitis (WEEV), which additionally incorporated a previously developed assay for WNV [149]. While a multiplex reverse transcription-quantitative polymerase chain reaction (RT-qPCR) has been developed for these viruses [172], it is not field-deployable [167]. The developed RT-LAMP targeted the non-structural protein 4 (nsP4) gene for WEEV, and the 3′ untranslated region (3′-UTR) for SEEV, and had a detection limit the same as the previously developed WNV RT-LAMP at 0.1 pfu per reaction [149,159]. Despite having a sensitivity marginally less than the previously developed RT-qPCR assay, both the SLEV and WEEV assays were performed in less than 30 min [159], and WNV RT-LAMP in under 17 min [149], supremely faster than the RT-qPCR assay. As this panel assay was performed on mosquitos in the field, it can be deployed as a large-scale surveillance program and as a rapid diagnostic technique [149,159].

Additionally, Foord et al. [26] developed a LAMP assay that was able to detect the conserved P-gene of Hendra virus before clinical signs appeared. This assay also compared utilizing a lateral flow device (LFD) to agarose gel electrophoresis for visual detection. While the LFD was not as sensitive in comparison to the gel, it was able to show results in five minutes, providing further confirmation of LAMP's field deployable abilities. Furthermore, the LAMP assay was able to detect additional positive results that was previously deemed "indeterminate" using a TaqMan assay. The authors suggest the simple procedure allow for LAMP to be employed in resource-poor environments. In addition, the capability of detecting positive cases prior to the onset of symptoms is ideal for critical situations, such as a Hendra virus outbreak, that require immediate results.

These developed assays show that both LAMP and RT-LAMP can be performed in-field as POC diagnostic technique. However, while various LAMP assays have been developed to detect viruses of concern to equines, both with high sensitivity and specificity, excluding the WNV RT-LAMP assay [149,159], none of these assays has been commercialized. The reasoning for this lack of commercially available assays remains unclear.

7. Application of iiPCR for Equine Viral Diseases

7.1. Principles of iiPCR

iiPCR is a recently developed assay involving an isothermal convective device [24]. The technique amplifies nucleic acids like PCR; however, it replaces the use of an expensive thermocycler with a simpler, portable, insulated device that consists of a copper ring attached to polycarbonate capillary tubes (R-tube™) underneath (Figure 1). The thermal convective device allows for reagents to proceed through gradient temperatures within the single tube [173], thus performing the required denaturation, annealing, and extension steps in a portable manner. Additionally, the insulation protects the assay from environmental influence, permitting its use in the field [24]. iiPCR has been analyzed as more sensitive than RT-PCR, achieving results within 1 h [155] through simple and cost-effective procedures [29,173,174].

7.2. Applications of iiPCR for Equine Viral Diseases

The iiPCR technique has been implemented for several equine viral diseases (Table 4). As seen in RT-LAMP, reverse-transcription has been integrated with iiPCR (iiRT-PCR) to detect RNA viruses through the generation of amplified cDNA [155].

Table 4. Current LAMP assays developed for equine viral diseases.

Disease	Type	Vector-Borne	Target Gene	Sample	Detection Limit	In-Field	Ref
Equine viral arteritis	ssRNA	No	ORF7	Horse—Tissue, semen	10 copies/rxn	Yes	[155]
Equine herpesvirus 3	dsDNA	No	gG	Horse—Perineal and genital swabs	6 copies/rxn	Yes	[175]
Equine herpesvirus myeloencephalopathy (EHM) caused by EHV-1	dsDNA	No	ORF3	Horse—Tissue	13 copies/rxn	Yes	[176]
Equine infectious anaemia	ssRNA	Yes—Horse and deer flies	5′ UTR Exon 1 of *tat* gene	Horse—Tissue	8 copies/rxn	Yes	[177]
Equine influenza (H3N8)	ssRNA	No	HA	Horse—Nasal swab	11 copies/rxn	Yes	[148]

Advancements of this technique have resulted in developing a portable machine that allows for automatic detection, termed POCKIT™ Nucleic Acid Analyzer by GeneReach USA (GeneReach USA, Lexington, MA, USA). This lightweight machine detects amplicons using hydrolysis technology recognizing fluorescent signals [173]. Carossino et al. [155] utilized this technology to develop a iiRT-PCR assay to detect EVA. This assay reported to have significant accuracy with a detection limit of 10 copies per reaction in one hour. Furthermore, compared to a previously developed RT-qPCR diagnostic test for EVA [178], the iiRT-PCR assay was ten-fold more sensitive. Therefore, this iiRT-PCR further exhibited the potential of future assays to be exploited in field for POC diagnostics.

Additionally, the assay did not encounter inhibition when using tissue samples that had been previously observed with the developed RT-qPCR assay [155]. The robustness of iiPCR and iiRT-PCR assays are advantageous as promising alternatives for diagnostic and control implementation [155,175,176]. However, despite numerous successful assays developed for equine infectious diseases, commercially available kits using the POCKIT™ Nucleic Acid Analyzer (GeneReach USA, Lexington, MA, USA) have been restricted to the aquaculture industry. Thus, for further traction of this technique and technology, commercialization should be made applicable to the equine industry.

8. Future Applications of LAMP and iiPCR for Equine Viral Diseases

Field deployable and POC assays for disease detection are becoming increasingly sought after [179–181], particularly for livestock and large animals to avoid transport-related stress and cost [182–184]. These portable diagnostic techniques allow for sampling and testing to take place pen-side or in the field without a laboratory [183], subsequently eliminating the transportation process and providing results in real-time for immediate treatment and control of infectious diseases [185,186]. Both LAMP and iiPCR are ideal for field-deployable diagnostics owing to their robustness, cost-efficiency, accessibility, and portable instruments, and are advantageous over conventional PCR assays (Table 5) [181,187,188].

Table 5. Comparison of conventional PCR to iiPCR and LAMP assays and procedures.

Properties	PCR	iiPCR	LAMP
Temperature	Cycles through 3 temperatures 55–95 °C	Constant temperature drives temperature gradient 15–30 °C	Constant temperature 60–65 °C
Equipment	Thermocycler	Specialized reaction tube Fluorescence-based detector	Heat source
Field-deployable	No	Yes	Yes
Reaction time	At least 90 min	≤60 min	<30 min
Sensitivity	Starts at nanograms	Starts at nanograms	Starts at femtograms
Specificity	Requires specific primer design Prone to errors	Requires specific primer design Prone to errors	Tolerates combination of primer designs
Visualization	Only through gel electrophoresis	Real-time available	Real-time available
Template prep	Requires purification	Requires purification	Tolerates impurities
Cost	$$$	$$	$

Both LAMP and iiPCR have portable machines that are lightweight (roughly 5 kg), robust, and only require an AC voltage or car battery to operate [189]. In addition, these machines are paving the way for newer diagnostic techniques, providing countless opportunities for alternative diagnostic technologies [181,188].

It should be noted that the feasibility of these assays is still dependent on sampling techniques and preparations [189]. Comparability, LAMP can tolerate sample impurities and inhibitory substances as it has been developed to eliminate nucleic acid extraction steps altogether [190,191], whereas iiPCR still requires nucleic acid purification [189]. In addition, inhibitors of PCR, can interfere with the assay results rendering them inaccurate and often false-negative outcomes. Therefore, sampling techniques that are in-field appropriate is a rapidly expanding area of research. Research groups have developed extraction systems that could isolate total nucleic acids through column-based methods. Despite these methods being described as user-friendly [192], contamination and degradation of RNA was still an issue, attributed to extensive manual handling throughout sample processing [189]. Thus, a field-deployable fully automated extraction system was developed by GeneReach, coined the Taco™ mini extraction system [193]. This machine can handle an array of samples, including more complex tissue and swab samples, to completely extract nucleic acids from up to eight samples concurrently. This magnetic beads-based tool is relatively inexpensive, compacted, and lightweight allowing for immediate use and practical storage after use [155,177,189,193]. However, this process adds an extra 45 min of processing to the assay, and while inexpensive, additional machinery is not suited to resource-poor facilities. The more intricate nucleic acid extraction requirements of iiPCR are hindering its POC application in comparison to the practical LAMP.

9. Conclusions

With an estimated value of USD $300 billion annually, involving more than 59 million domestic horses and 1.6 million full-time employees, it is essential to protect the global equine industry from disease outbreaks. Despite strict worldwide biosecurity procedures, the threat of a viral outbreak, including zoonotic diseases, remains imminent. Due to the increasing amount of national and international movement and subsequent dense housing of horse populations, the spread of viral diseases could be rapid and devastating, particularly with asymptomatic carriers. Current "gold-standard" diagnostic techniques, such as serological and molecular technology, remain prominent within the industry; however, they come with several drawbacks that limit their use, particularly in resource-poor settings. Newer isothermal techniques, such as LAMP and iiPCR, allow for rapid diagnosis and offer the opportunity to be field-deployable. However, further research is required to ultimately eliminate laborious procedures, particularly in nucleic acid extraction. While LAMP has been developed to tolerate sample impurities and does not require extraction steps, iiPCR continues to rely on extra machinery to provide an automated extraction technique. Nevertheless, the use of these methodologies remains advantageous over traditional methods for POC testing, based on their rapidity, sensitivity, specificity, and inexpensiveness. Thus, there is strong reasoning to develop new diagnostics using isothermal technology as alternatives to traditional techniques for rapid disease identification and quick implementation of control measures.

Author Contributions: Conceptualization, A.K. and T.B.; data curation, A.K.; writing—original draft preparation, A.K.; writing—review and editing, A.K. and T.B.; supervision, T.B.; funding acquisition, A.K. and. T.B. All authors have read and agreed to the published version of the manuscript.

Funding: This research was supported by Cooperative Research Centres Project (CRC-P) awarded to Geneworks and La Trobe University. A.K. is supported by an La Trobe Industry PhD scholarship and Defence Science Institute, an initiative of the State Government of Victoria.

Institutional Review Board Statement: Not applicable.

Acknowledgments: Authors would like Gemma Zerna for her valuable advice during the preparation of the manuscript.

Conflicts of Interest: The authors declare no conflict of interest.

References

1. Murray, G.; Munstermann, S.; Lam, K. Benefits and Challenges Posed by the Worldwide Expansion of Equestrian Events—New Standards for the Population of Competition Horses and Equine Disease Free Zones (EDFZ) in Countries. In Proceedings of the 81st General Session World Organisation for Animal Health, Paris, France, 26–31 May 2013.
2. Littiere, T.O.; Castro, G.H.F.; Rodriguez, M.D.P.R.; Bonafé, C.M.; Magalhães, A.F.B.; Faleiros, R.R.; Vieira, J.I.G ; Santos, C.G.; Verardo, L.L. Identification and Functional Annotation of Genes Related to Horses' Performance: From GWAS to Post-GWAS. *Animals* **2020**, *10*, 1173. [CrossRef]
3. Paillot, R. Special Issue "Equine Viruses": Old "Friends" and New Foes? *Viruses* **2020**, *12*, 153. [CrossRef]
4. FAOSTAT. Production Statistics of the Food Agriculture Orginization of The United States. Available online: http //www.fao.org/faostat/en/#data/QA (accessed on 3 June 2021).
5. Cross, P. *Global Horse Statistics Internal 02 2019*; HiPoint Agro Bedding Corp: Gulep, ON, Canada, 2019.
6. Hatcher, F. *Equine Industry Scoping Report*; Regional Development Australia—Far South Coast Nowra: Nowra, NSW, Australia, 2013.
7. Pritchard, J.C.; Lindberg, A.C.; Main, D.C.; Whay, H.R. Assessment of the welfare of working horses, mules and donkeys, using health and behaviour parameters. *Prev. Vet. Med.* **2005**, *69*, 265–283. [CrossRef]
8. McManus, P.; Albrecht, G.; Graham, R. *The Global Horseracing Industry: Social, Economic, Environmental and Ethical Perspectives. The Global Horseracing Industry: Social, Economic, Environmental and Ethical Perspectives*; Routledge: London, UK, 2012; pp. 1–244. [CrossRef]
9. IFHA. *Annual Report 2018*; International Federation of Horseracing Authorities: Boulogne, France, 2018.
10. Narayan, P.K.; Smyth, R.L. The race that stops the nation: The demand for the Melbourne Cup. *Econ. Rec.* **2004**, *80*, 193–207. [CrossRef]
11. Hardy, G.L.P. *Measurement of Economic Impact of Australian Thoroughbred Breeding Industry 18/046*; NSW: Wagga Wagga, Australia, 2019.
12. Attoui, H.; Mohd Jaafar, F. Zoonotic and emerging orbivirus infections. *Rev. Sci. Tech.* **2015**, *34*, 353–361. [CrossRef]
13. Sack, A.; Oladunni, F.S.; Gonchigoo, B.; Chambers, T.M.; Gray, G.C. Zoonotic Diseases from Horses: A Systematic Review. *Vector Borne Zoonotic. Dis.* **2020**, *20*, 484–495. [CrossRef]
14. Yuen, K.Y.; Bielefeldt-Ohmann, H. Ross River Virus Infection: A Cross-Disciplinary Review with a Veterinary Perspective. *Pathogens* **2021**, *10*, 357. [CrossRef]
15. Smyth, G.B.; Dagley, K.; Tainsh, J. Insights into the economic consequences of the 2007 equine influenza outbreak in Australia. *Aust. Vet. J.* **2011**, *89* (Suppl. 1), 151–158. [CrossRef] [PubMed]

16. Prasad, M.; Brar, B.; Shah, I.; Ranjan, K.; Lambe, U.; Manimegalai, M.; Vashisht, B.; Khurana, S.; Prasad, G. Biotechnological tools for diagnosis of equine infectious diseases. *J. Exp. Biol. Agric. Sci.* **2016**, *4*, S161–S181. [CrossRef]
17. Weese, J.S. Infection control and biosecurity in equine disease control. *Equine Vet. J.* **2014**, *46*, 654–660. [CrossRef]
18. Slovis, N.M.; Browne, N.; Bozorgmanesh, R. Point-of-Care Diagnostics in Equine Practice. *Vet. Clin. N. Am. Equine Pract* **2020**, *36*, 161–171. [CrossRef]
19. Desmettre, P. Diagnosis and prevention of equine infectious diseases: Present status, potential, and challenges for the future. *Adv. Vet. Med.* **1999**, *41*, 359–377. [CrossRef]
20. Zhang, X.; Lowe, S.B.; Gooding, J.J. Brief review of monitoring methods for loop-mediated isothermal amplification (LAMP). *Biosens. Bioelectron.* **2014**, *61*, 491–499. [CrossRef]
21. Notomi, T.; Mori, Y.; Tomita, N.; Kanda, H. Loop-mediated isothermal amplification (LAMP): Principle, features, and future prospects. *J. Microbiol.* **2015**, *53*, 1–5. [CrossRef]
22. Nagura-Ikeda, M.; Imai, K.; Tabata, S.; Miyoshi, K.; Murahara, N.; Mizuno, T.; Horiuchi, M.; Kato, K.; Imoto, Y.; Iwata, M.; et al. Clinical Evaluation of Self-Collected Saliva by Quantitative Reverse Transcription-PCR (RT-qPCR), Direct RT-qPCR, Reverse Transcription-Loop-Mediated Isothermal Amplification, and a Rapid Antigen Test To Diagnose COVID-19. *J. Clin. Microbiol.* **2020**, *58*, e01438-20. [CrossRef]
23. Notomi, T.; Okayama, H.; Masubuchi, H.; Yonekawa, T.; Watanabe, K.; Amino, N.; Hase, T. Loop-mediated isothermal amplification of DNA. *Nucleic Acids Res.* **2000**, *28*, E63. [CrossRef]
24. Chang, H.-F.G.; Tsai, Y.-L.; Tsai, C.-F.; Lin, C.-K.; Lee, P.-Y.; Teng, P.-H.; Su, C.; Jeng, C.-C. A thermally baffled device for highly stabilized convective PCR. *Biotechnol. J.* **2012**, *7*, 662–666. [CrossRef]
25. Alhassan, A.; Thekisoe, O.M.; Yokoyama, N.; Inoue, N.; Motloang, M.Y.; Mbati, P.A.; Yin, H.; Katayama, Y.; Anzai, T.; Sugimoto, C.; et al. Development of loop-mediated isothermal amplification (LAMP) method for diagnosis of equine piroplasmosis. *Vet. Parasitol.* **2007**, *143*, 155–160. [CrossRef]
26. Foord, A.J.; Middleton, D.; Heine, H.G. Hendra virus detection using Loop-Mediated Isothermal Amplification. *J. Virol. Methods* **2012**, *181*, 93–96. [CrossRef] [PubMed]
27. Bath, C.; Scott, M.; Sharma, P.M.; Gurung, R.B.; Phuentshok, Y.; Pefanis, S.; Colling, A.; Singanallur Balasubramanian, N.; Firestone, S.M.; Ungvanijban, S.; et al. Further development of a reverse-transcription loop-mediated isothermal amplification (RT-LAMP) assay for the detection of foot-and-mouth disease virus and validation in the field with use of an internal positive control. *Transbound. Emerg. Dis.* **2020**, *67*, 2494–2506. [CrossRef]
28. Chua, K.H.; Lee, P.C.; Chai, H.C. Development of insulated isothermal PCR for rapid on-site malaria detection. *Malar. J.* **2016**, *15*, 134. [CrossRef] [PubMed]
29. Tsai, Y.L.; Wang, H.T.; Chang, H.F.; Tsai, C.F.; Lin, C.K.; Teng, P.H.; Su, C.; Jeng, C.C.; Lee, P.Y. Development of TaqMan probe-based insulated isothermal PCR (iiPCR) for sensitive and specific on-site pathogen detection. *PLoS ONE* **2012**, *7*, e45278. [CrossRef]
30. Schemann, K.; Taylor, M.R.; Toribio, J.A.; Dhand, N.K. Horse owners' biosecurity practices following the first equine influenza outbreak in Australia. *Prev. Vet. Med.* **2011**, *102*, 304–314. [CrossRef]
31. OIE. *Manual of Diagnostic Tests and Vaccines for Terrestrial Animals, 2018*; World Organisation for Animal Health (OIE): Paris, France, 2019.
32. OIE. Notifable Animal Diseases. Available online: https://www.oie.int/en/what-we-do/animal-health-and-welfare/animal-diseases/ (accessed on 8 June 2021).
33. Mellor, P.S.; Hamblin, C. African horse sickness. *Vet. Res.* **2004**, *35*, 445–466. [CrossRef] [PubMed]
34. Mellor, P.S. African horse sickness: Transmission and epidemiology. *Vet. Res.* **1993**, *24*, 199–212. [PubMed]
35. Sellers, R.F.; Pedgley, D.E.; Tucker, M.R. Possible spread of African horse sickness on the wind. *J. Hyg.* **1977**, *79*, 279–298. [CrossRef]
36. Maclachlan, N.J.; Guthrie, A.J. Re-emergence of bluetongue, African horse sickness, and other orbivirus diseases. *Vet. Res.* **2010**, *41*, 35. [CrossRef]
37. Gao, H.; Bie, J.; Wang, H.; Zheng, J.; Gao, X.; Xiao, J.; Wang, H. *Modelling High.-Risk Areas for African Horse Sickness Occurrence in Mainland China Along Southeast*; Authorea: Hobken, NJ, USA, 2020.
38. King, S.; Rajko-Nenow, P.; Ashby, M.; Frost, L.; Carpenter, S.; Batten, C. Outbreak of African horse sickness in Thailand, 2020. *Transbound. Emerg. Dis.* **2020**, *67*, 1764–1767. [CrossRef]
39. Lu, G.; Pan, J.; Ou, J.; Shao, R.; Hu, X.; Wang, C.; Li, S. African horse sickness: Its emergence in Thailand and potential threat to other Asian countries. *Transbound. Emerg. Dis.* **2020**, *67*, 1751–1753. [CrossRef]
40. Dennis, S.J.; Meyers, A.E.; Hitzeroth, I.I.; Rybicki, E.P. African Horse Sickness: A Review of Current Understanding and Vaccine Development. *Viruses* **2019**, *11*, 844. [CrossRef]
41. Robin, M.; Page, P.; Archer, D.; Baylis, M. African horse sickness: The potential for an outbreak in disease-free regions and current disease control and elimination techniques. *Equine. Vet. J.* **2016**, *48*, 659–669. [CrossRef] [PubMed]
42. Forrester, N.L.; Kenney, J.L.; Deardorff, E.; Wang, E.; Weaver, S.C. Western Equine Encephalitis submergence: Lack of evidence for a decline in virus virulence. *Virology* **2008**, *380*, 170–172. [CrossRef] [PubMed]
43. Bergren, N.A.; Auguste, A.J.; Forrester, N.L.; Negi, S.S.; Braun, W.A.; Weaver, S.C. Western equine encephalitis virus: Evolutionary analysis of a declining alphavirus based on complete genome sequences. *J. Virol.* **2014**, *88*, 9260–9267. [CrossRef] [PubMed]

44. Robb, L.; Hartman, D.; Rice, L.; DeMaria, J.; Borland, E.; Bergren, N.; Kading, R. Continued Evidence of Decline in the Enzootic Activity of Western Equine Encephalitis Virus in Colorado. *J. Med Entomol.* **2018**, *56*, 584–588. [CrossRef]
45. Bergren, N.A.; Haller, S.; Rossi, S.L.; Seymour, R.L.; Huang, J.; Miller, A.L.; Bowen, R.A.; Hartman, D.A.; Brault, A.C ; Weaver, S.C. "Submergence" of Western equine encephalitis virus: Evidence of positive selection argues against genetic drift and fitness reductions. *PLoS Pathog* **2020**, *16*, e1008102. [CrossRef] [PubMed]
46. Calisher, C.H.; Monath, T.P.; Mitchell, C.J.; Sabattini, M.S.; Cropp, C.B.; Kerschner, J.; Hunt, A.R.; Lazuick, J.S. Arbovirus investigations in Argentina, 1977-1980. III. Identification and characterization of viruses isolated, including new subtypes of western and Venezuelan equine encephalitis viruses and four new bunyaviruses (Las Maloyas, Resistencia, Barranqueras, and Antequera). *Am. J. Trop Med. Hyg.* **1985**, *34*, 956–965.
47. Avilés, G.; Sabattini, M.S.; Mitchell, C.J. Transmission of western equine encephalomyelitis virus by Argentine Aedes albifasciatus (Diptera: Culicidae). *J. Med. Entomol.* **1992**, *29*, 850–853. [CrossRef]
48. Stromberg, Z.R.; Fischer, W.; Bradfute, S.B.; Kubicek-Sutherland, J.Z.; Hraber, P. Vaccine Advances against Venezuelan, Eastern, and Western Equine Encephalitis Viruses. *Vaccines* **2020**, *8*, 273. [CrossRef]
49. Anderson, B.A. Focal neurologic signs in western equine encephalitis. *Can. Med. Assoc. J.* **1984**, *130*, 1019–1021. [PubMed]
50. Reed, D.S.; Larsen, T.; Sullivan, L.J.; Lind, C.M.; Lackemeyer, M.G.; Pratt, W.D.; Parker, M.D. Aerosol exposure to western equine encephalitis virus causes fever and encephalitis in cynomolgus macaques. *J. Infect. Dis.* **2005**, *192*, 1173–1182. [CrossRef]
51. Pellegrini-Masini, A.; Livesey, L.C. Meningitis and encephalomyelitis in horses. *Vet. Clin. N. Am. Equine. Pract.* **2006**, *22*, 553–589. [CrossRef]
52. CDC. Arboviral disease—United States, 1994. *MMWR Morb. Mortal. Wkly. Rep.* **1995**, *44*, 641–644.
53. Bleck, T. *Arboviruses Affecting the Central Nervous System. Goldman's Cecil Medicine*, 24th ed.; Elsevier Saunders: Philadelphia, PA, USA, 2011; Volume 2, pp. 2161–2168. [CrossRef]
54. Lambert, A.J.; Martin, D.A.; Lanciotti, R.S. Detection of North American eastern and western equine encephalitis viruses by nucleic acid amplification assays. *J. Clin. Microbiol.* **2003**, *41*, 379–385. [CrossRef] [PubMed]
55. Sellon, D.C. Equine infectious anemia. *Vet. Clin. N. Am. Equine. Pract.* **1993**, *9*, 321–336. [CrossRef]
56. Craigo, J.K.; Montelaro, R.C. Lessons in AIDS vaccine development learned from studies of equine infectious, anemia virus infection and immunity. *Viruses* **2013**, *5*, 2963–2976. [CrossRef] [PubMed]
57. Cursino, A.E.; Vilela, A.P.P.; Franco-Luiz, A.P.M.; de Oliveira, J.G.; Nogueira, M.F.; Júnior, J.P.A.; de Aguiar, D.M.; Kroon, E.G. Equine infectious anemia virus in naturally infected horses from the Brazilian Pantanal. *Arch. Virol.* **2018**, *163*, 2385–2394. [CrossRef] [PubMed]
58. Lupulovic, D.; Savić, S.; Gaudaire, D.; Berthet, N.; Grgić, Ž.; Matović, K.; Deshiere, A.; Hans, A. Identification and genetic characterization of equine infectious anemia virus in Western Balkans. *BMC Vet. Res.* **2021**, *17*, 168. [CrossRef]
59. Cook, R.F.; Leroux, C.; Issel, C.J. Equine infectious anemia and equine infectious anemia virus in 2013: A review. *Vet. Microbiol.* **2013**, *167*, 181–204. [CrossRef]
60. Cruz, F.; Fores, P.; Ireland, J.; Moreno, M.A.; Newton, R. Freedom from equine infectious anaemia virus infection in Spanish Purebred horses. *Vet. Rec. Open* **2015**, *2*, e000074. [CrossRef] [PubMed]
61. Wang, H.-N.; Rao, D.; Fu, X.-Q.; Hu, M.-M.; Dong, J.-G. Equine infectious anemia virus in China. *Oncotarget* **2017**, *9*, 1356–1364. [CrossRef]
62. Issel, C.J.; Coggins, L. Equine infectious anemia: Current knowledge. *J. Am. Vet. Med. Assoc.* **1979**, *174*, 727–733.
63. Bolfa, P.; Barbuceanu, F.; Leau, S.E.; Leroux, C. Equine infectious anaemia in Europe: Time to re-examine the efficacy of monitoring and control protocols? *Equine Vet. J.* **2016**, *48*, 140–142. [CrossRef]
64. Espasandin, A.G.; Cipolini, M.F.; Forletti, A.; Díaz, S.; Soto, J.; Martínez, D.E.; Storani, C.A.; Monzón, N.M.; Beltrame, J.I.; Lucchesi, E.; et al. Comparison of serological techniques for the diagnosis of equine infectious Anemia in an endemic area of Argentina. *J. Virol. Methods* **2021**, *291*, 114101. [CrossRef]
65. McConnico, R.S.; Issel, C.J.; Cook, S.J.; Cook, R.F.; Floyd, C.; Bisson, H. Predictive methods to define infection with equine infectious anemia virus in foals out of reactor mares. *J. Equine Vet. Sci.* **2000**, *20*, 387–392. [CrossRef]
66. Sack, A.; Cullinane, A.; Daramragchaa, U.; Chuluunbaatar, M.; Gonchigoo, B.; Gray, G.C. Equine Influenza Virus—A Neglected, Reemergent Disease Threat. *Emerg. Infect. Dis.* **2019**, *25*, 1185–1191. [CrossRef]
67. van Maanen, C.; Cullinane, A. Equine influenza virus infections: An update. *Vet. Q.* **2002**, *24*, 79–94. [CrossRef]
68. Taylor, M.R.; Agho, K.E.; Stevens, G.J.; Raphael, B. Factors influencing psychological distress during a disease epidemic: Data from Australia's first outbreak of equine influenza. *BMC Public Health* **2008**, *8*, 347. [CrossRef] [PubMed]
69. Cowled, B.; Ward, M.P.; Hamilton, S.; Garner, G. The equine influenza epidemic in Australia: Spatial and temporal descriptive analyses of a large propagating epidemic. *Prev. Vet. Med.* **2009**, *92*, 60–70. [CrossRef] [PubMed]
70. Elton, D.; Bryant, N. Facing the threat of equine influenza. *Equine Vet. J.* **2011**, *43*, 250–258. [CrossRef] [PubMed]
71. Mena, J.; Brito, B.; Moreira, R.; Tadich, T.; González, I.; Cruces, J.; Ortega, R.; van Bakel, H.; Rathnasinghe, R.; Pizarro-Lucero, J.; et al. Reemergence of H3N8 Equine Influenza A virus in Chile, 2018. *Transbound. Emerg. Dis.* **2018**, *65*, 1408–1415. [CrossRef]
72. Cullinane, A.; Elton, D.; Mumford, J. Equine influenza—Surveillance and control. *Influenza Other Respir. Viruses* **2010**, *4*, 339–344. [CrossRef]
73. Balasuriya, U.B.; Carossino, M. Reproductive effects of arteriviruses: Equine arteritis virus and porcine reproductive and respiratory syndrome virus infections. *Curr. Opin. Virol.* **2017**, *27*, 57–70. [CrossRef] [PubMed]

74. Del Piero, F. Equine viral arteritis. *Vet. Pathol.* **2000**, *37*, 287–296. [CrossRef]
75. Timoney, P.J.; McCollum, W.H. Equine viral arteritis—Epidemiology and control. *J. Equine Vet. Sci.* **1988**, *8*, 54–59. [CrossRef]
76. Timoney, P.J.; McCollum, W.H. Equine viral arteritis. *Vet. Clin. N. Am. Equine Pract.* **1993**, *9*, 295–309. [CrossRef]
77. Balasuriya, U.B.R.; Carossino, M.; Timoney, P.J. Equine viral arteritis: A respiratory and reproductive disease of significant economic importance to the equine industry. *Equine Vet. Educ.* **2018**, *30*, 497–512. [CrossRef]
78. Ruiz-Saenz, J. Equine Viral Arteritis: Epidemiological and intervention perspectives. *Revista Colombiana Ciencias Pecuarias* **2010**, *23*, 501.
79. Pronost, S.; Pitel, P.H.; Miszczak, F.; Legrand, L.; Marcillaud-Pitel, C.; Hamon, M.; Tapprest, J.; Balasuriya, U.B.; Freymuth, F.; Fortier, G. Description of the first recorded major occurrence of equine viral arteritis in France. *Equine Vet. J.* **2010**, *42*, 713–720. [CrossRef]
80. Crabb, B.S.; Studdert, M.J. Equine herpesviruses 4 (equine rhinopneumonitis virus) and 1 (equine abortion virus). *Adv. Virus Res.* **1995**, *45*, 153–190. [CrossRef] [PubMed]
81. Dunowska, M. A review of equid herpesvirus 1 for the veterinary practitioner. Part B: Pathogenesis and epidemiology. *N. Z. Vet. J.* **2014**, *62*, 179–188. [CrossRef] [PubMed]
82. Balasuriya, U.B.; Crossley, B.M.; Timoney, P.J. A review of traditional and contemporary assays for direct and indirect detection of Equid herpesvirus 1 in clinical samples. *J. Vet. Diagn Investig.* **2015**, *27*, 673–687. [CrossRef] [PubMed]
83. Fitzpatrick, D.R.; Studdert, M.J. Immunologic relationships between equine herpesvirus type 1 (equine abortion virus) and type 4 (equine rhinopneumonitis virus). *Am. J. Vet. Res.* **1984**, *45*, 1947–1952.
84. Patel, J.R.; Heldens, J. Equine herpesviruses 1 (EHV-1) and 4 (EHV-4)—Epidemiology, disease and immunoprophylaxis: A brief review. *Vet. J.* **2005**, *170*, 14–23. [CrossRef] [PubMed]
85. Lunn, D.P.; Davis-Poynter, N.; Flaminio, M.J.; Horohov, D.W.; Osterrieder, K.; Pusterla, N.; Townsend, H.G. Equine herpesvirus-1 consensus statement. *J. Vet. Intern. Med.* **2009**, *23*, 450–461. [CrossRef]
86. Ma, G.; Azab, W.; Osterrieder, N. Equine herpesviruses type 1 (EHV-1) and 4 (EHV-4)—Masters of co-evolution and a constant threat to equids and beyond. *Vet. Microbiol.* **2013**, *167*, 123–134. [CrossRef] [PubMed]
87. Kydd, J.H.; Townsend, H.G.; Hannant, D. The equine immune response to equine herpesvirus-1: The virus and its vaccines. *Vet. Immunol. Immunopathol.* **2006**, *111*, 15–30. [CrossRef] [PubMed]
88. Powell, D.G. Viral respiratory disease of the horse. *Vet. Clin. N. Am. Equine Pract* **1991**, *7*, 27–52. [CrossRef]
89. Reed, S.M.; Toribio, R.E. Equine herpesvirus 1 and 4. *Vet. Clin. N. Am. Equine Pract.* **2004**, *20*, 631–642. [CrossRef] [PubMed]
90. Oladunni, F.S.; Horohov, D.W.; Chambers, T.M. EHV-1: A Constant Threat to the Horse Industry. *Front. Microbiol.* **2019**, *10*, 140–142. [CrossRef] [PubMed]
91. Field, H.; de Jong, C.; Melville, D.; Smith, C.; Smith, I.; Broos, A.; Kung, Y.H.N.; McLaughlin, A.; Zeddeman, A. Hendra virus infection dynamics in Australian fruit bats. *PLoS ONE* **2011**, *6*, e28678. [CrossRef]
92. Khusro, A.; Aarti, C.; Pliego, A.B.; Cipriano-Salazar, M. Hendra Virus Infection in Horses: A Review on Emerging Mystery Paramyxovirus. *J. Equine Vet. Sci.* **2020**, *91*, 103149. [CrossRef]
93. Mahalingam, S.; Herrero, L.J.; Playford, E.G.; Spann, K.; Herring, B.; Rolph, M.S.; Middleton, D.; McCall, B.; Field, H.; Wang, L.F. Hendra virus: An emerging paramyxovirus in Australia. *Lancet Infect. Dis.* **2012**, *12*, 799–807. [CrossRef]
94. Playford, G.; McCall, B.; Smith, G.; Slinko, V.; Allen, G.; Smith, I.; Moore, F.; Taylor, C.; Kung, N.; Field, H. Human Hendra Virus Encephalitis Associated with Equine Outbreak, Australia, 2008. *Emerg. Infect. Dis.* **2010**, *16*, 219–223. [CrossRef]
95. Mendez, D.H.; Judd, J.; Speare, R. Unexpected result of Hendra virus outbreaks for veterinarians, Queensland, Australia. *Emerg. Infect. Dis.* **2012**, *18*, 83–85. [CrossRef] [PubMed]
96. Hii, C.; Dhand, N.K.; Toribio, J.-A.L.M.L.; Taylor, M.R.; Wiethoelter, A.; Schembri, N.; Sawford, K.; Kung, N.; Moloney, B.; Wright, T.; et al. Information delivery and the veterinarian-horse owner relationship in the context of Hendra virus in Australia. *Prev. Vet. Med.* **2020**, *179*, 104988. [CrossRef] [PubMed]
97. Whitley, R.J.; Gnann, J.W. Viral encephalitis: Familiar infections and emerging pathogens. *Lancet* **2002**, *359*, 507–513. [CrossRef]
98. Mackenzie, J.S.; Johansen, C.A.; Ritchie, S.A.; van den Hurk, A.F.; Hall, R.A. Japanese encephalitis as an emerging virus: The emergence and spread of Japanese encephalitis virus in Australasia. *Curr. Top. Microbiol. Immunol.* **2002**, *267*, 49–73. [CrossRef] [PubMed]
99. Lam, K.; Ellis, T.; Williams, D.; Lunt, R.; Daniels, P.; Watkins, K.; Riggs, C. Japanese encephalitis in a racing Thoroughbred gelding in Hong Kong. *Vet. Rec.* **2005**, *157*, 168–173. [CrossRef] [PubMed]
100. Mansfield, K.L.; Hernández-Triana, L.M.; Banyard, A.C.; Fooks, A.R.; Johnson, N. Japanese encephalitis virus infection, diagnosis and control in domestic animals. *Vet. Microbiol.* **2017**, *201*, 85–92. [CrossRef]
101. van den Hurk, A.F.; Ritchie, S.A.; Mackenzie, J.S. Ecology and geographical expansion of Japanese encephalitis virus. *Annu. Rev. Entomol.* **2009**, *54*, 17–35. [CrossRef] [PubMed]
102. Russell, R.C. Ross River virus: Ecology and distribution. *Annu. Rev. Entomol.* **2002**, *47*, 1–31. [CrossRef]
103. Johnson, B.J.; Robbins, A.; Gyawali, N.; Ong, O.; Loader, J.; Murphy, A.K.; Hanger, J.; Devine, G.J. The environmental and ecological determinants of elevated Ross River Virus exposure in koalas residing in urban coastal landscapes. *Sci. Rep.* **2021**, *11*, 4419. [CrossRef]

104. Liu, X.; Mutso, M.; Cherkashchenko, L.; Zusinaite, E.; Herrero, L.J.; Doggett, S.L.; Haniotis, J.; Merits, A.; Herring, B.L.; Taylor, A.; et al. Identification of Natural Molecular Determinants of Ross River Virus Type I Interferon Modulation. *J. Virol.* **2020**, *94*, e01788-01719. [CrossRef] [PubMed]
105. Hall, N.L.; Barnes, S.; Canuto, C.; Nona, F.; Redmond, A.M. Climate change and infectious diseases in Australia's Torres Strait Islands. *Aust. N. Z. J. Public Health* **2021**, *45*, 122–128. [CrossRef]
106. Azuolas, J.K. Ross River Virus Disease of Horses. *Aust. Equine Vet.* **1988**, *16*, 3.
107. El-Hage, C.M.; Bamford, N.J.; Gilkerson, J.R.; Lynch, S.E. Ross River Virus Infection of Horses: Appraisal of Ecological and Clinical Consequences. *J. Equine Vet. Sci.* **2020**, *93*, 103143. [CrossRef]
108. Chapman, G.E.; Baylis, M.; Archer, D.; Daly, J.M. The challenges posed by equine arboviruses. *Equine Vet. J.* **2018**, *50*, 436–445. [CrossRef]
109. Harley, D.; Sleigh, A.; Ritchie, S. Ross River virus transmission, infection, and disease: A cross-disciplinary review. *Clin. Microbiol. Rev.* **2001**, *14*, 909–932. [CrossRef]
110. Campbell, G.L.; Marfin, A.A.; Lanciotti, R.S.; Gubler, D.J. West Nile virus. *Lancet Infect. Dis.* **2002**, *2*, 519–529. [CrossRef]
111. Petersen, L.R.; Brault, A.C.; Nasci, R.S. West Nile virus: Review of the literature. *JAMA* **2013**, *310*, 308–315. [CrossRef]
112. Hayes, E.B.; Komar, N.; Nasci, R.S.; Montgomery, S.P.; O'Leary, D.R.; Campbell, G.L. Epidemiology and transmission dynamics of West Nile virus disease. *Emerg. Infect. Dis.* **2005**, *11*, 1167–1173. [CrossRef]
113. Sharifi, Z.; Mahmoodian Shooshtari, M.; Talebian, A. A study of West Nile virus infection in Iranian blood donors. *Arch. Iran. Med.* **2010**, *13*, 1–4.
114. Zeller, H.G.; Schuffenecker, I. West Nile virus: An overview of its spread in Europe and the Mediterranean basin in contrast to its spread in the Americas. *Eur. J. Clin. Microbiol. Infect. Dis.* **2004**, *23*, 147–156. [CrossRef]
115. Cantile, C.; Di Guardo, G.; Eleni, C.; Arispici, M. Clinical and neuropathological features of West Nile virus equine encephalomyelitis in Italy. *Equine Vet. J.* **2000**, *32*, 31–35. [CrossRef]
116. Bertram, F.M.; Thompson, P.N.; Venter, M. Epidemiology and Clinical Presentation of West Nile Virus Infection in Horses in South Africa, 2016–2017. *Pathogens* **2020**, *10*, 20. [CrossRef] [PubMed]
117. Salazar, P.; Morley, P.; Wilmot, D.; Steffen, D.; Cunningham, W.; Salman, M. Outcome of equids with clinical signs of West Nile virus infection and factors associated with death. *J. Am. Vet. Med Assoc.* **2004**, *225*, 267–274. [CrossRef] [PubMed]
118. Bosco-Lauth, A.M.; Bowen, R.A. West Nile Virus: Veterinary Health and Vaccine Development. *J. Med. Entomol.* **2019**, *56*, 1463–1466. [CrossRef]
119. Ulbert, S. West Nile virus vaccines—Current situation and future directions. *Hum. Vaccin Immunother.* **2019**, *15*, 2337–2342. [CrossRef] [PubMed]
120. Gubler, D.J.; Campbell, G.L.; Nasci, R.; Komar, N.; Petersen, L.; Roehrig, J.T. West Nile virus in the United States: Guidelines for detection, prevention, and control. *Viral. Immunol.* **2000**, *13*, 469–475. [CrossRef]
121. Kramer, L.D.; Li, J.; Shi, P.-Y. West Nile virus. *Lancet Neurol.* **2007**, *6*, 171–181. [CrossRef]
122. Sambri, V.; Capobianchi, M.R.; Cavrini, F.; Charrel, R.; Donoso-Mantke, O.; Escadafal, C.; Franco, L.; Gaibani, P.; Gould, E.A.; Niedrig, M.; et al. Diagnosis of west nile virus human infections: Overview and proposal of diagnostic protocols considering the results of external quality assessment studies. *Viruses* **2013**, *5*, 2329–2348. [CrossRef]
123. Rizzoli, A.; Jimenez-Clavero, M.A.; Barzon, L.; Cordioli, P.; Figuerola, J.; Koraka, P.; Martina, B.; Moreno, A.; Nowotny, N.; Pardigon, N.; et al. The challenge of West Nile virus in Europe: Knowledge gaps and research priorities. *Eurosurveillance* **2015**, *20*. [CrossRef]
124. Eidson, M.; Komar, N.; Sorhage, F.; Nelson, R.; Talbot, T.; Mostashari, F.; McLean, R.; West Nile Virus Avian Mortality Surveillance, G. Crow deaths as a sentinel surveillance system for West Nile virus in the northeastern United States, 1999. *Emerg. Infect. Dis.* **2001**, *7*, 615–620. [CrossRef] [PubMed]
125. Reed, K.D.; Meece, J.K.; Henkel, J.S.; Shukla, S.K. Birds, migration and emerging zoonoses: West nile virus, lyme disease, influenza A and enteropathogens. *Clin. Med. Res.* **2003**, *1*, 5–12. [CrossRef] [PubMed]
126. Selvey, L.A.; Donnelly, J.A.; Lindsay, M.D.; PottumarthyBoddu, S.; D'Abrera, V.C.; Smith, D.W. Ross River virus infection surveillance in the Greater Perth Metropolitan area–has there been an increase in cases in the winter months? *Commun. Dis. Intell. Q. Rep.* **2014**, *38*, E114–E122. [PubMed]
127. Kinsley, R.; Scott, S.D.; Daly, J.M. Controlling equine influenza: Traditional to next generation serological assays. *Vet. Microbiol.* **2016**, *187*, 15–20. [CrossRef]
128. Zimmerman, K.L.; Crisman, M.V. Diagnostic equine serology. *Vet. Clin. N. Am. Equine Pract.* **2008**, *24*, 311–334. [CrossRef] [PubMed]
129. Ranjan, K.; Prasad, M.; Prasad, G. Application of Molecular and Serological Diagnostics in Veterinary Parasitology. *J. Adv. Parasitol.* **2016**, *2*, 80–99. [CrossRef]
130. Cappelli, K.; Capomaccio, S.; Cook, F.R.; Felicetti, M.; Marenzoni, M.L.; Coppola, G.; Verini-Supplizi, A.; Coletti, M.; Passamonti, F. Molecular detection, epidemiology, and genetic characterization of novel European field isolates of equine infectious anemia virus. *J. Clin. Microbiol.* **2011**, *49*, 27–33. [CrossRef]
131. Jaffer, O.; Abdishakur, F.; Hakimuddin, F.; Riya, A.; Wernery, U.; Schuster, R.K. A comparative study of serological tests and PCR for the diagnosis of equine piroplasmosis. *Parasitol. Res.* **2010**, *106*, 709–713. [CrossRef]

132. Mott, J.; Rikihisa, Y.; Zhang, Y.; Reed, S.M.; Yu, C.Y. Comparison of PCR and culture to the indirect fluorescent-antibody test for diagnosis of Potomac horse fever. *J. Clin. Microbiol.* **1997**, *35*, 2215–2219. [CrossRef]
133. Varrasso, A.; Dynon, K.; Ficorilli, N.; Hartley, C.A.; Studdert, M.J.; Drummer, H.E. Identification of equine herpesviruses 1 and 4 by polymerase chain reaction. *Aust. Vet. J.* **2001**, *79*, 563–569. [CrossRef]
134. Sachse, K. Specificity and performance of PCR detection assays for microbial pathogens. *Mol. Biotechnol.* **2004**, *26*, 61–80. [CrossRef]
135. Wilson, I.G. Inhibition and facilitation of nucleic acid amplification. *Appl. Environ. Microbiol.* **1997**, *63*, 3741–3751. [CrossRef] [PubMed]
136. Cohen, N.D.; Martin, L.J.; Simpson, R.B.; Wallis, D.E.; Neibergs, H.L. Comparison of polymerase chain reaction and microbiological culture for detection of salmonellae in equine feces and environmental samples. *Am. J. Vet. Res.* **1996**, *57*, 780–786. [PubMed]
137. Furr, M.; MacKay, R.; Granstrom, D.; Schott, H., 2nd; Andrews, F. Clinical diagnosis of equine protozoal myeloencephalitis (EPM). *J. Vet. Intern. Med.* **2002**, *16*, 618–621. [CrossRef]
138. Paxson, J. Evaluating polymerase chain reaction-based tests for infectious pathogens. *Compend. Equine* **2008**, *3*, 308–320.
139. Pusterla, N.; Madigan, J.E.; Leutenegger, C.M. Real-time polymerase chain reaction: A novel molecular diagnostic tool for equine infectious diseases. *J. Vet. Intern. Med.* **2006**, *20*, 3–12. [CrossRef]
140. Diallo, I.S.; Hewitson, G.; Wright, L.; Rodwell, B.J.; Corney, B.G. Detection of equine herpesvirus type 1 using a real-time polymerase chain reaction. *J. Virol. Methods* **2006**, *131*, 92–98. [CrossRef]
141. Kim, C.M.; Blanco, L.B.; Alhassan, A.; Iseki, H.; Yokoyama, N.; Xuan, X.; Igarashi, I. Diagnostic real-time PCR assay for the quantitative detection of Theileria equi from equine blood samples. *Vet. Parasitol.* **2008**, *151*, 158–163. [CrossRef]
142. Fernández-Pinero, J.; Fernández-Pacheco, P.; Rodríguez, B.; Sotelo, E.; Robles, A.; Arias, M.; Sánchez-Vizcaíno, J.M. Rapid and sensitive detection of African horse sickness virus by real-time PCR. *Res. Vet. Sci.* **2009**, *86*, 353–358. [CrossRef]
143. Aeschbacher, S.; Santschi, E.; Gerber, V.; Stalder, H.P.; Zanoni, R.G. Development of a real-time RT-PCR for detection of equine influenza virus. *Schweiz Arch. Tierheilkd* **2015**, *157*, 191–201. [CrossRef]
144. Zanoli, L.M.; Spoto, G. Isothermal amplification methods for the detection of nucleic acids in microfluidic devices. *Biosensors* **2012**, *3*, 18–43. [CrossRef] [PubMed]
145. Bodulev, O.L.; Sakharov, I.Y. Isothermal Nucleic Acid Amplification Techniques and Their Use in Bioanalysis. *Biochemistry* **2020**, *85*, 147–166. [CrossRef] [PubMed]
146. Obande, G.A.; Banga Singh, K.K. Current and Future Perspectives on Isothermal Nucleic Acid Amplification Technologies for Diagnosing Infections. *Infect. Drug Resist.* **2020**, *13*, 455–483. [CrossRef]
147. Vincent, M.; Xu, Y.; Kong, H. Helicase-dependent isothermal DNA amplification. *EMBO Rep.* **2004**, *5*, 795–800. [CrossRef]
148. Balasuriya, U.B.R.; Lee, P.-Y.A.; Tiwari, A.; Skillman, A.; Nam, B.; Chambers, T.M.; Tsai, Y.-L.; Ma, L.-J.; Yang, P.-C.; Chang, H.-F.G.; et al. Rapid detection of equine influenza virus H3N8 subtype by insulated isothermal RT-PCR (iiRT-PCR) assay using the POCKIT™ Nucleic Acid Analyzer. *J. Virol. Methods* **2014**, *207*, 66–72. [CrossRef]
149. Parida, M.; Posadas, G.; Inoue, S.; Hasebe, F.; Morita, K. Real-time reverse transcription loop-mediated isothermal amplification for rapid detection of West Nile virus. *J. Clin. Microbiol.* **2004**, *42*, 257–263. [CrossRef]
150. Dean, F.B.; Hosono, S.; Fang, L.; Wu, X.; Faruqi, A.F.; Bray-Ward, P.; Sun, Z.; Zong, Q.; Du, Y.; Du, J.; et al. Comprehensive human genome amplification using multiple displacement amplification. *Proc. Natl. Acad. Sci. USA* **2002**, *99*, 5261–5266. [CrossRef] [PubMed]
151. Compton, J. Nucleic acid sequence-based amplification. *Nature* **1991**, *350*, 91–92. [CrossRef]
152. Fire, A.; Xu, S.Q. Rolling replication of short DNA circles. *Proc. Natl. Acad. Sci. USA* **1995**, *92*, 4641–4645. [CrossRef] [PubMed]
153. Piepenburg, O.; Williams, C.H.; Stemple, D.L.; Armes, N.A. DNA detection using recombination proteins. *PLoS Biol.* **2006**, *4*, e204. [CrossRef] [PubMed]
154. Walker, G.T.; Fraiser, M.S.; Schram, J.L.; Little, M.C.; Nadeau, J.G.; Malinowski, D.P. Strand displacement amplification–an isothermal, in vitro DNA amplification technique. *Nucleic Acids Res.* **1992**, *20*, 1691–1696. [CrossRef] [PubMed]
155. Carossino, M.; Lee, P.Y.; Nam, B.; Skillman, A.; Shuck, K.M.; Timoney, P.J.; Tsai, Y.L.; Ma, L.J.; Chang, H.F.; Wang, H.T.; et al. Development and evaluation of a reverse transcription-insulated isothermal polymerase chain reaction (RT-iiPCR) assay for detection of equine arteritis virus in equine semen and tissue samples using the POCKIT™ system. *J. Virol. Methods* **2016**, *234*, 7–15. [CrossRef]
156. Nagamine, K.; Watanabe, K.; Ohtsuka, K.; Hase, T.; Notomi, T. Loop-mediated isothermal amplification reaction using a nondenatured template. *Clin. Chem.* **2001**, *47*, 1742–1743. [CrossRef] [PubMed]
157. Mori, Y.; Notomi, T. Loop-mediated isothermal amplification (LAMP): A rapid, accurate, and cost-effective diagnostic method for infectious diseases. *J. Infect. Chemother.* **2009**, *15*, 62–69. [CrossRef]
158. Parida, M.; Sannarangaiah, S.; Dash, P.K.; Rao, P.V.; Morita, K. Loop mediated isothermal amplification (LAMP): A new generation of innovative gene amplification technique; perspectives in clinical diagnosis of infectious diseases. *Rev. Med. Virol.* **2008**, *18*, 407–421. [CrossRef]
159. Wheeler, S.S.; Ball, C.S.; Langevin, S.A.; Fang, Y.; Coffey, L.L.; Meagher, R.J. Surveillance for Western Equine Encephalitis, St. Louis Encephalitis, and West Nile Viruses Using Reverse Transcription Loop-Mediated Isothermal Amplification. *PLoS ONE* **2016**, *11*, e0147962. [CrossRef] [PubMed]

160. Silva, S.; Pardee, K.; Pena, L. Loop-Mediated Isothermal Amplification (LAMP) for the Diagnosis of Zika Virus: A Review. *Viruses* **2019**, *12*, 19. [CrossRef]
161. Dukes, J.P.; King, D.P.; Alexandersen, S. Novel reverse transcription loop-mediated isothermal amplification for rapid detection of foot-and-mouth disease virus. *Arch. Virol.* **2006**, *151*, 1093–1106. [CrossRef]
162. Han, Q.; Zhang, S.; Liu, D.; Yan, F.; Wang, H.; Huang, P.; Bi, J.; Jin, H.; Feng, N.; Cao, Z.; et al. Development of a Visible Reverse Transcription-Loop-Mediated Isothermal Amplification Assay for the Detection of Rift Valley Fever Virus. *Front. Microbiol.* **2020**, *11*, 590732. [CrossRef] [PubMed]
163. Fowler, V.L.; Howson, E.L.A.; Flannery, J.; Romito, M.; Lubisi, A.; Agüero, M.; Mertens, P.; Batten, C.A.; Warren, H.R.; Castillo-Olivares, J. Development of a Novel Reverse Transcription Loop-Mediated Isothermal Amplification Assay for the Rapid Detection of African Horse Sickness Virus. *Transbound. Emerg. Dis.* **2017**, *64*, 1579–1588. [CrossRef]
164. Nemoto, M.; Tsujimura, K.; Yamanaka, T.; Kondo, T.; Matsumura, T. Loop-mediated isothermal amplification assays for detection of Equid herpesvirus 1 and 4 and differentiating a gene-deleted candidate vaccine strain from wild-type Equid herpesvirus 1 strains. *J. Vet. Diagn. Investig.* **2010**, *22*, 30–36. [CrossRef] [PubMed]
165. Han, J.; Zhang, H.; Tang, Q.; Liu, T.C.; Xu, J.; Li, L.R.; Yu, G.; Wang, Y.; Yoa, X.P.; Yang, Z.X. Preliminary study of LAMP method for the detection of equine infectious anemia virus. In Proceedings of the 2017 2nd International Conference on Biomedical and Biological Engineering (BBE2017), Guilin, China, 26–28 May 2017; pp. 492–499.
166. Nemoto, M.; Yamanaka, T.; Bannai, H.; Tsujimura, K.; Kondo, T.; Matsumura, T. Development and evaluation of a reverse transcription loop-mediated isothermal amplification assay for H3N8 equine influenza virus. *J. Virol. Methods* **2011**, *178*, 239–242. [CrossRef]
167. Nemoto, M.; Yamanaka, T.; Bannai, H.; Tsujimura, K.; Kondo, T.; Matsumura, T. Development of a reverse transcription loop-mediated isothermal amplification assay for H7N7 equine influenza virus. *J. Vet. Med. Sci.* **2012**, *74*, 929–931. [CrossRef]
168. Nemoto, M.; Morita, Y.; Niwa, H.; Bannai, H.; Tsujimura, K.; Yamanaka, T.; Kondo, T. Rapid detection of equine coronavirus by reverse transcription loop-mediated isothermal amplification. *J. Virol. Methods* **2015**, *215-216*, 13–16. [CrossRef] [PubMed]
169. Vissani, M.A.; Becerra, M.L.; Olguín Perglione, C.; Tordoya, M.S.; Miño, S.; Barrandeguy, M. Neuropathogenic and non-neuropathogenic genotypes of Equid Herpesvirus type 1 in Argentina. *Vet. Microbiol.* **2009**, *139*, 361–364. [CrossRef]
170. Waters, R.A.; Fowler, V.L.; Armson, B.; Nelson, N.; Gloster, J.; Paton, D.J.; King, D.P. Preliminary validation of direct detection of foot-and-mouth disease virus within clinical samples using reverse transcription loop-mediated isothermal amplification coupled with a simple lateral flow device for detection. *PLoS ONE* **2014**, *9*, e105630. [CrossRef] [PubMed]
171. Howson, E.L.A.; Armson, B.; Madi, M.; Kasanga, C.J.; Kandusi, S.; Sallu, R.; Chepkwony, E.; Siddle, A.; Martin, P.; Wood, J.; et al. Evaluation of Two Lyophilized Molecular Assays to Rapidly Detect Foot-and-Mouth Disease Virus Directly from Clinical Samples in Field Settings. *Transbound. Emerg. Dis.* **2017**, *64*, 861–871. [CrossRef]
172. Brault, A.C.; Fang, Y.; Reisen, W.K. Multiplex qRT-PCR for the Detection of Western Equine Encephalomyelitis, St. Louis Encephalitis, and West Nile Viral RNA in Mosquito Pools (Diptera: Culicidae). *J. Med. Entomol.* **2015**, *52*, 491–499. [CrossRef]
173. Tsai, Y.L.; Lin, Y.C.; Chou, P.H.; Teng, P.H.; Lee, P.Y. Detection of white spot syndrome virus by polymerase chain reaction performed under insulated isothermal conditions. *J. Virol. Methods* **2012**, *181*, 134–137. [CrossRef]
174. Wilkes, R.P.; Anis, E.; Dunbar, D.; Lee, P.A.; Tsai, Y.L.; Lee, F.C.; Chang, H.G.; Wang, H.T.; Graham, E.M. Rapid and sensitive insulated isothermal PCR for point-of-need feline leukaemia virus detection. *J. Feline Med. Surg.* **2018**, *20*, 362–369. [CrossRef]
175. Vissani, M.A.; Tordoya, M.S.; Tsai, Y.L.; Lee, P.Y.A.; Shen, Y.H.; Lee, F.C.; Wang, H.T.T.; Parreño, V.; Barrandeguy, M. On-site detection of equid alphaherpesvirus 3 in perineal and genital swabs of mares and stallions. *J. Virol. Methods* **2018**, *257*, 29–32. [CrossRef]
176. Balasuriya, U.B.; Lee, P.A.; Tsai, Y.L.; Tsai, C.F.; Shen, Y.H.; Chang, H.G.; Skillman, A.; Wang, H.T.; Pronost, S.; Zhang, Y. Translation of a laboratory-validated equine herpesvirus-1 specific real-time PCR assay into an insulated isothermal polymerase chain reaction (iiPCR) assay for point-of-need diagnosis using POCKIT™ nucleic acid analyzer. *J. Virol. Methods* **2017**, *241*, 58–63. [CrossRef]
177. Cook, R.F.; Barrandeguy, M.; Lee, P.A.; Tsai, C.F.; Shen, Y.H.; Tsai, Y.L.; Chang, H.G.; Wang, H.T.; Balasuriya, U.B.R. Rapid detection of equine infectious anaemia virus nucleic acid by insulated isothermal RT-PCR assay to aid diagnosis under field conditions. *Equine Vet. J.* **2019**, *51*, 489–494. [CrossRef] [PubMed]
178. Balasuriya, U.B.R.; Leutenegger, C.M.; Topol, J.B.; McCollum, W.H.; Timoney, P.J.; MacLachlan, N.J. Detection of equine arteritis virus by real-time TaqMan® reverse transcription-PCR assay. *J. Virol. Methods* **2002**, *101*, 21–28. [CrossRef]
179. Tomlinson, J. In-field diagnostics using loop-mediated isothermal amplification. *Methods Mol. Biol.* **2013**, *938*, 291–300. [CrossRef] [PubMed]
180. Schar, D.; Padungtod, P.; Tung, N.; O'Leary, M.; Kalpravidh, W.; Claes, F. New frontiers in applied veterinary point-of-capture diagnostics: Toward early detection and control of zoonotic influenza. *Influenza Other Respir. Viruses* **2019**, *13*, 618–621. [CrossRef] [PubMed]
181. Moehling, T.J.; Choi, G.; Dugan, L.C.; Salit, M.; Meagher, R.J. LAMP Diagnostics at the Point-of-Care: Emerging Trends and Perspectives for the Developer Community. *Expert Rev. Mol. Diagn* **2021**, *21*, 43–61. [CrossRef]
182. Minka, N.S.; Ayo, J.O. Effects of loading behaviour and road transport stress on traumatic injuries in cattle transported by road during the hot-dry season. *Livestock Sci.* **2007**, *107*, 91–95. [CrossRef]

183. Nayak, S.; Blumenfeld, N.R.; Laksanasopin, T.; Sia, S.K. Point-of-Care Diagnostics: Recent Developments in a Connected Age. *Anal. Chem.* **2017**, *89*, 102–123. [CrossRef]
184. Bhatt, N.; Singh, N.P.; Mishra, A.; Rajneesh, R.; Jamwal, S. A detailed review of transportation stress in livestock and its mitigation techniques. *Int. J. Livest. Res.* **2021**, *1*, 30–41. [CrossRef]
185. Vashist, S.K.: Luppa, P.B.; Yeo, L.Y.; Ozcan, A.; Luong, J.H.T. Emerging Technologies for Next-Generation Point-of-Care Testing. *Trends Biotechnol.* **2015**, *33*, 692–705. [CrossRef]
186. St John, A.; Price, C.P. Existing and Emerging Technologies for Point-of-Care Testing. *Clin. Biochem. Rev.* **2014**, *35*, 155–167.
187. Kemleu, S.; Guelig, D.; Eboumbou Moukoko, C.; Essangui, E.; Diesburg, S.; Mouliom, A.; Melingui, B.; Manga, J.; Donkeu, C.; Epote, A.; et al. A Field-Tailored Reverse Transcription Loop-Mediated Isothermal Assay for High Sensitivity Detection of Plasmodium falciparum Infections. *PLoS ONE* **2016**, *11*, e0165506. [CrossRef]
188. Sathish Kumar, T.; Radhika, K.; Joseph Sahaya Rajan, J.; Makesh, M.; Alavandi, S.V.; Vijayan, K.K. Closed-tube field-deployable loop-mediated isothermal amplification (LAMP) assay based on spore wall protein (SWP) for the visual detection of Enterocytozoon hepatopenaei (EHP). *J. Invertebr. Pathol.* **2021**, *183*, 107624. [CrossRef]
189. Ambagala, A.; Fisher, M.; Goolia, M.; Nfon, C.; Furukawa-Stoffer, T.; Ortega Polo, R.; Lung, O. Field-Deployable Reverse Transcription-Insulated Isothermal PCR (RT-iiPCR) Assay for Rapid and Sensitive Detection of Foot-and-Mouth Disease Virus. *Transbound. Emerg. Dis.* **2017**, *64*, 1610–1623. [CrossRef]
190. Dudley, D.M.; Newman, C.M.; Weiler, A.M.; Ramuta, M.D.; Shortreed, C.G.; Heffron, A.S.; Accola, M.A.; Rehrauer, W.M.; Friedrich, T.C.; O'Connor, D.H. Optimizing direct RT-LAMP to detect transmissible SARS-CoV-2 from primary nasopharyngeal swab samples. *PLoS ONE* **2020**, *15*, e0244882. [CrossRef]
191. Dao Thi, V.L.; Herbst, K.; Boerner, K.; Meurer, M.; Kremer, L.P.; Kirrmaier, D.; Freistaedter, A.; Papagiannidis, D.; Galmozzi, C.; Stanifer, M.L.: et al. A colorimetric RT-LAMP assay and LAMP-sequencing for detecting SARS-CoV-2 RNA in clinical samples. *Sci. Transl. Med.* **2020**, *12*. [CrossRef]
192. Ambagala, A.; Pahari, S.; Fisher, M.; Lee, P.A.; Pasick, J.; Ostlund, E.N.; Johnson, D.J.; Lung, O. A Rapid Field-Deployable Reverse Transcription-Insulated Isothermal Polymerase Chain Reaction Assay for Sensitive and Specific Detection of Bluetongue Virus. *Transbound. Emerg. Dis.* **2017**, *64*, 476–486. [CrossRef]
193. Carossino, M.: Li, Y.; Lee, P.-Y.A.; Tsai, C.-F.; Chou, P.-H.; Williams, D.; Skillman, A.; Frank Cook, R.; Brown, G.; Chang, H.-F.G.; et al. Evaluation of a field-deployable reverse transcription-insulated isothermal PCR for rapid and sensitive on-site detection of Zika virus. *BMC Infect. Dis.* **2017**, *17*, 778. [CrossRef]

 MDPI

Review

Equine Encephalosis Virus

Sharon Tirosh-Levy [1,2,*] and Amir Steinman [1]

1 Koret School of Veterinary Medicine, The Robert H. Smith Faculty of Agriculture, Food and Environment, The Hebrew University of Jerusalem, Rehovot 7610001, Israel; amirst@savion.huji.ac.il
2 Division of Parasitology, Kimron Veterinary Institute, Beit Dagan 50200, Israel
* Correspondence: sharontirosh@gmail.com

Simple Summary: Equine encephalosis (EE) is a febrile disease of horses caused by EE virus (EEV) and transmitted by *Culicoides* midges. This virus was first isolated from a horse in South Africa in 1967 and until 2008 was believed to be restricted to southern Africa. In 2008–2009, isolation of EEV in an outbreak reported from Israel demonstrated the emergence of this pathogen into new niches. Indeed, further testing revealed that EEV had already spread outside of South Africa since 2001. Although EEV normally does not cause severe clinical disease, it should be considered important since it may indicate the possible spread of other related, much more pathogenic viruses, such as African horse sickness virus (AHSV). The spread of EEV from South Africa to central Africa, the Middle East, and India is an example of the possible emergence of new pathogens in new niches and should be a reminder not to limit the differential diagnoses list when facing a possible outbreak or a cluster of undiagnosed clinical cases. This review summarizes current knowledge regarding EEV structure, pathogenesis, clinical significance, and epidemiology.

Abstract: Equine encephalosis (EE) is an arthropod-borne, noncontagious, febrile disease of horses. It is caused by EE virus (EEV), an *Orbivirus* of the Reoviridae family transmitted by *Culicoides*. Within the EEV serogroup, seven serotypes (EEV-1–7) have been identified to date. This virus was first isolated from a horse in South Africa in 1967 and until 2008 was believed to be restricted to southern Africa. In 2008–2009, isolation of EEV in an outbreak reported from Israel demonstrated the emergence of this pathogen into new niches. Indeed, testing in retrospect sera samples revealed that EEV had already been circulating outside of South Africa since 2001. Although EEV normally does not cause severe clinical disease, it should be considered important since it may indicate the possible spread of other related, much more pathogenic viruses, such as African horse sickness virus (AHSV). The spread of EEV from South Africa to central Africa, the Middle East and India is an example of the possible emergence of new pathogens in new niches, as was seen in the case of West Nile virus, and should be a reminder not to limit the differential list when facing a possible outbreak or a cluster of clinical cases. This review summarizes current knowledge regarding EEV structure, pathogenesis, clinical significance, and epidemiology.

Keywords: equine encephalosis virus; EEV; horse; epidemiology; clinical disease; control; *Culicoides*

Citation: Tirosh-Levy, S.; Steinman, A. Equine Encephalosis Virus. *Animals* **2022**, *12*, 337. https://doi.org/10.3390/ani12030337

Academic Editor: Serena Montagnaro

Received: 10 January 2022
Accepted: 26 January 2022
Published: 29 January 2022

1. Introduction

Equine encephalosis (EE) is an arthropod-borne, noncontagious, febrile disease of horses. It was first described over a century ago by Theiler, as a mild form of African horse sickness (AHS), under the name "equine ephemeral fever" [1], and was first isolated in 1967 in South Africa from a thoroughbred mare named Cascara that died following febrile nervous disease [2]. The disease is caused by equine encephalosis virus (EEV), an *Orbivirus* of the Reoviridae family, closely related to several other important pathogenic and emerging viruses affecting livestock, including bluetongue virus (BTV), African horse sickness virus (AHSV), and epizootic hemorrhagic disease virus (EHDV), all transmitted by *Culicoides* species [3].

The clinical significance of EEV is probably low, as it usually manifests as mild, transient, febrile disease, which is rarely fatal [3,4]. The risk factors for infection and vector species are similar to those of AHSV, and both viruses usually circulate in the same areas [5–7]. Although EEV was considered to be endemic only in southern Africa, reports of its presence in other areas have been accumulating for over a decade [4,8–11]. These reports coincide with the spread and emergence of other *Orbiviruses* in Asia and Europe due to the combination of animal transport and climate changes leading to changes in *Culicoides* habitat [3,7]. Since EEV is less pathogenic, it may be more easily introduced into new areas and may serve as an indicator of the potential spread of other more clinically important *Orbiviruses,* including AHSV [12].

2. Etiology

EEV is an arbovirus of the genus *Orbivirus,* subfamily Sedoreovirinae, and family Reoviridae, transmitted by hematophagous *Culicoides* spp. [13]. The genus *Orbivirus* consists of over 20 serogroups and is the largest genus within the family Reoviridae [13]. Within the EEV serogroup, seven serotypes (EEV-1–7) have been identified to date [14].

The viral genome consists of 10 segments of linear double-stranded RNA (dsRNA), surrounded by three layers of capsid proteins, forming a double-layered core particle or inner capsid, surrounded by an outer capsid layer. Virus particles are 60–80 nm in diameter, have icosahedral symmetry, and appear spherical in shape [13,15] (Figure 1).

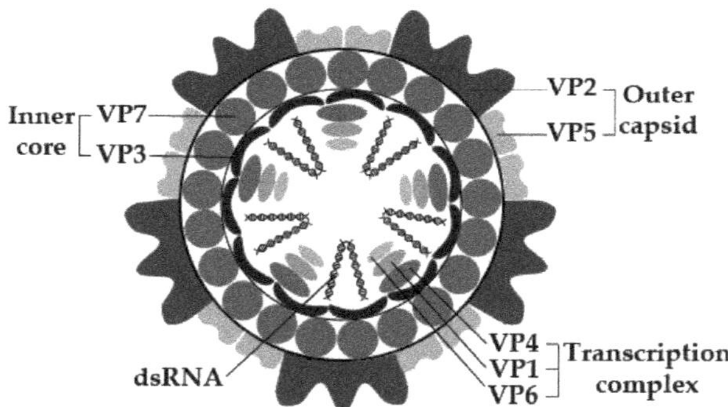

Figure 1. The molecular structure of EEV, according to electron microscopy and molecular studies of EEV, and closely related *Orbiviruses*. Visualization was based on the work of [13,16] and created using Adobe Illustrator 25.4.1© (Adobe Inc., Mountain View, CA, USA).

Similar to other *Orbiviruses,* EEV has seven structural proteins (VP1–7) and four non-structural proteins (NS1–3, NS3a) [13,15,17] (Figure 1, Table 1). The structural proteins include four major capsid proteins (VP2, VP3, VP5, and VP7) and three minor proteins (VP1, VP4, and VP6), with molecular mass (Mr), which ranges between 36,000 and 120,000 [15,18]. The double-layered inner core comprises minor proteins VP1, VP4 and VP6, enwrapped by the major proteins VP3 and VP7. The minor proteins have enzymatic activities involved in viral replication and transcription [13,15,17]. The outer capsid layer comprises two proteins, VP2 and VP5, which are involved in cell attachment and penetration (along with VP7) and possibly in the determination of virulence [13,15,17,18]. Both VP2 and P7 are immunodominant, with VP2 being highly variable and determining EEV serotype [14,15,19].

Table 1. Genome segments and protein encoded by each segment of EEV (similar to other *Orbiviruses*).

Genome Segment	Protein	Function
Seg-1	VP1	RNA-dependent RNA polymerase
Seg-2	VP2	Protein of the outer layer of the outer capsid, involves in cell attachment, most variable, determines serotype
Seg-3	VP3	Innermost protein capsid shell
Seg-4	VP4	Capping enzyme
Seg-5	NS1	Forms tubules of unknown function
Seg-6	VP5	Inner layer of outer capsid, involves in cell penetration
Seg-7	VP7	Protein of the outer core surface, involves in cell entry, immunodominant
Seg-8	NS2	Inclusion body matrix protein
Seg-9	VP6/VP6A	Helicase
Seg-10	NS3/NS3A	Membrane protein, involves in cell exit, variable

The 5′- (5′-GUU(U/A)) and 3′- (A(U/A/G)(A/U/C)GUUAC-3′) terminal sequences of gene segments are conserved for all segments within the EEV serogroup [13,15]. Each genomic segment has a single open reading frame (ORF) (Table 1). Seg-9 and Seg-10 mRNAs are translated from either of two in-frame AUG codons (VP6/VP6A, NS3/NS3A); however, the significance of these different translation products is unclear [13] (Table 1). Genomic segments 3, 5, and 9 are serogroup specific and highly conserved between EEV serotypes and, therefore, may be used to distinguish between EEV and other closely related *Orbiviruses* [18,19]. Seg-2, encoding VP2, shows sequence variations that correlate with the virus-serotype [19]. The smallest viral genome segment, Seg-10, encodes NS3/NS3A, which mediates viral release from infected cells and may determine virulence and vector competence. The EEV NS3 gene and protein have a higher level of variation than in other *Orbiviruses*, and phylogenetic studies identified two distinct clusters that correspond with the geographical distribution of different species of *Culicoides* vectors [20]. The ability of *Orbiviruses* to undergo gene reassortment within a single serogroup has resulted in the absence of correlation between virus serotypes and sequence variations in other genomic segments [15,19].

The pathogenesis of EEV infection and replication is similar to other *Orbiviruses*, and involves: (1) cell attachment and penetration, which occurs soon after inoculation (and involve VP2); (2) uncoating and formation of replicative complexes after entry into the cell, the virus is enclosed in endosomes, in which the outer capsid is removed (involving VP5), resulting in the release of transcriptionally active core particles into the cytoplasm; (3) formation of cellular tubules (consisting NS1 and involving VP3 and VP7) with unknown functions that possibly interact with the cellular cytoskeleton and formation of virus inclusion bodies (containing a different combination of virus particles, with main involvement of NS2); and (4) movement of virus and its release from the cell surface (involving NS3/NS3A [15,21].

Molecular characterization on EEV was based mostly on the variable proteins/genomic sequences VP2 (Seg-2) and NS3 (Seg-10). Seven EEV serotypes have been characterized in South Africa based on the variable protein VP2 (and corresponding genomic sequence Seg-2). The serotypes were assigned numeric values based on the alphabetic order of the location in South Africa where the reference strain originated, namely: EEV-1 (Bryanston, 1976), EEV-2 (Cascara, 1967), EEV-3 (Gamil, 1971), EEV-4 (Kaalplaas, 1974), EEV-5 (Kyalami, 1974), EEV-6 (Potchefstroom, 1991), and EEV-7 (E21/20, 2000) [14]. Phylogenetic analysis of the NS3 gene of South African EEV isolates grouped them into two clusters, which differ by up to 16.7% in amino acid sequence identity. Cluster A included serotypes EEV-1, 2, 4, and 7, and cluster B included serotypes EEV-3, 5, and 6, corresponding to the geographical distribution of the isolates [20].

3. Epidemiology

Until 2008, EE had only been reported in southern Africa. Since its first isolation in 1967, additional isolations and epidemiological surveys demonstrated the widespread circulation of EEV in horses and donkeys in South Africa and identified seven serotypes [5,14,22,23]. During the 1990s, EEV seropositivity has also been described in donkeys and zebras in South Africa and neighboring Botswana, Kenya, and Namibia [23–26]. In 2008, EEV was isolated in Israel during an outbreak of febrile disease in horses [9,27], and further serological studies demonstrated that the virus has been circulating in Israel since 2001 [28] and that it is also present in neighboring Palestinian Authority and Jordan [29]. Following the initial report from Israel demonstrating EEV outside southern Africa, sero-epidemiological studies revealed the virus is endemic in eastern and western African countries, including Ghana (2010), Gambia (2009), Ethiopia (2008), and Zimbabwe (since 1999) [8,10]. However, none of 120 horses sampled in Morocco (during 2008) were found seropositive for EEV, suggesting that the Sahara Desert may serve as a geographical barrier to the spread of the virus [10]. In addition, EEV was isolated in India from a horse that died in 2008 following febrile disease and identified with the help of next-generation sequencing [11]. This clinical case was the only fatal case during an outbreak of febrile disease on the farm, and since that was the first report of EEV in India, its prevalence or spread in the area is yet unknown [11] (Figure 2).

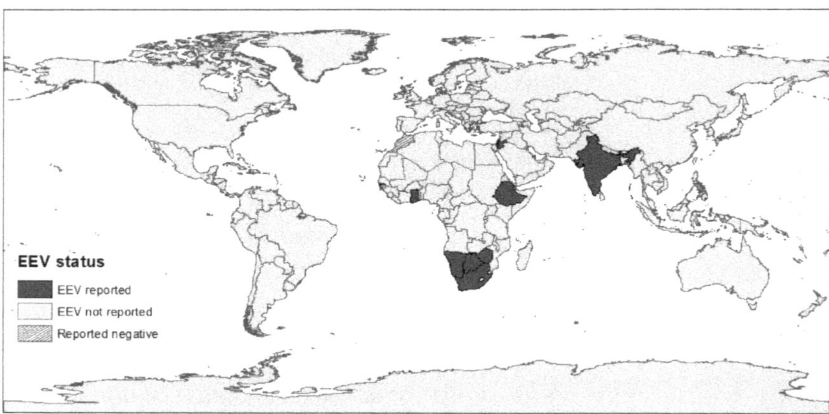

Figure 2. The global distribution of EEV, according to serological and molecular studies. The map was constructed using ArcMap (ArcGIS Desktop 10.6.1, Esri Inc.©, Redlands, CA, USA).

EEV is endemic in Africa and in the Middle East, with seroprevalence ranging between 60% and 100% in Gambia, Ghana, Ethiopia, Israel, South Africa, and Zimbabwe [5,6,8,10,29]. Studies from South Africa and Israel revealed fluctuations in the annual seroprevalence and incidence of EEV, which may be influenced by weather, climate, herd immunity, and the distribution of *Culicoides* vector species [5,6,22,23]. Spatial and temporal studies from South Africa showed that EEV seroprevalence and the abundance of specific EEV serotypes differ between geographical provinces, but also between districts within the same province [5,22,23]. All EEV serotypes have been identified in all South African provinces, but the relative abundance of each serotype varied between areas and even between farms. In each area and season, there was usually one predominant circulating serotype (with demonstrated seroconversion), while others were only isolated sporadically [5,14,22]. However, similar serotypes were identified in horses and *Culicoides* in the same area [5]. It has been demonstrated that individual horses can be simultaneously seropositive to several serotypes, which indicates that there is no sufficient immunological cross-protection between serotypes against infection [5,14,22,30].

Sequence analyses of genomic segments provide additional information to serotype classification (which is based on serologic reaction to VP2 protein). Sequencing of the VP2 serotype 4 gene of isolates from Gambia and Israel found them unique to South African isolates and grouped them together, suggesting a common source outside of South Africa [10,27]. Characterization of NS3 genomic sequences of serotype 1 from an outbreak in Western Cape (1999) encoded identical proteins, which indicates a high level of conservation during the outbreak [20]. Although NS3 did not cluster according to geographic location [20,31], close phylogenetic relationships were found between EEV isolates from horses and from *Culicoides* during the same period and area [32], supporting the correlation found between the geographical distribution of EEV serotypes in horses and *Culicoides* and NS3 phylogenetic clusters [5]. Recent analysis on the full sequences of all 10 genome segments of 17 EEV isolates of all serotypes revealed widespread reassortment in EEV strains, with unique segments that may be associated with geographic location [32]. For example, the EEV isolate from India was classified as serotype 1 according to its VP2 sequence but was similar to serotype 6 according to its NS3 and VP1 sequences and had a unique combination of other segments than any of the South African isolates [11,32]. Field isolates from clinical cases had as little as 81.6% amino acid similarity to their corresponding serotype reference strains. This limited similarity may suggest genetic drift, which may possibly lead to immune evasion [32].

EEV is biologically transmitted by *Culicoides* biting midges, of which *C. imicola* and *C. bolitinos* have been shown to play an important role in South Africa [5,33,34]. The *Culicoides* genus includes over 1400 described species that inhabit a wide range of habitats. Only a small proportion of these species are known vectors of *Orbiviruses*; however, most studies only focus on certain species [35,36]. In South Africa, EEV was identified in *Culicoides* blood pools of several species [5,31,33,34,37], with high recovery rates (VRR) immediately following feeding (81.9%) and virus survival and multiplication demonstrated in five of 19 species tested following incubation for 10 days [5]. The mean levels of viral replication differed significantly between EEV serotypes and *Culicoides* species, suggesting that certain species have served as better vectors to specific serotypes [5] (Table 2).

Table 2. Survival of EEV serotypes in various *Culicoides* species following 10 days incubation at 23.5 °C after membrane feeding of infected blood [5].

Culicoides spp.	EEV Serotype Survival
C. imicola	EEV-1 >> 4 > 2,5 > 3 > 6
C. bolitinos	EEV-2 > 1 >> 4 > 6
C. leucostictus	EEV-1 > 2
C. magnus	EEV-1
C. zuluensis	EEV-2

In South Africa, *C. imicola* and *C. bolitinos* are the most abundant and widespread species and were identified as important vectors of EEV and other *Orbiviruses* such as BTV and AHSV. The higher vector competence of these two species for EEV-1 correlates with the high field recovery rate of this serotype from horses in South Africa [5,14,22]. The differences in prevalence and rate of exposure to individual serotypes within and between regions in South Africa may be attributed to the differences in their spatial and temporal distribution of certain *Culicoides* species, in combination with the differences in the competence of these vectors to specific serotypes. *C. imicola* is the main species in the northern regions of South Africa, which correspond with NS3 gene cluster B, while *C. bolitinos* is more abundant in the southern districts, corresponding with NS3 gene cluster A [5,20,38]. In addition, the low EEV seroprevalence in the Western Cape Province in South Africa could be attributed to the lower abundance of *C. imicola* in this region [5].

The epidemiology of *Culicoides*-borne diseases is often complex and involves multiple vectors and hosts within a geographical region [35]. *Culicoides* species have a worldwide distribution (except for Antarctica and New Zealand). The success of *Culicoides* to serve as vectors is related to their population size and means of dispersal, which are highly influenced by climate and weather [39]. Therefore, climatic differences, which affect the spatial and temporal distribution of *Culicoides*, may influence the prevalence of certain EEV serotypes in specific geographical areas or during an outbreak [5]. Although EEV is endemic in South Africa and normally has minimal clinical significance, local outbreaks are sometimes reported [14,30]. Long-term epidemiological studies from South Africa, Israel, and Zimbabwe demonstrated fluctuation in the rate of infection between years [6,8,22]. In South Africa, the annual seroprevalence in yearling foals ranged between 3.6% and 34.7% [22] and varied between EEV serotypes [14,22]. These fluctuations may reflect changes in vector distribution. Seasonal drought followed by heavy rainfall had been shown to increase the chance of arboviral diseases, including the EE outbreak in Israel in 2008 [6] and AHS outbreaks in South Africa [22,33,40]. This association was mainly explained by an effect of water deficit on the environment, altering the relationships between vectors and hosts (as water sources may be more available in farms), but might also be the result of changes in the vectorial capacity of the insects, inferred by drought [41]. Since several important *Orbiviruses* share the same *Culicoides* vectors, their spread and outbreaks often coincide. Several studies demonstrated similar infection patterns of EEV and AHSV in equids, with usually higher prevalence of EEV than of AHSV in both hosts and vectors [8,23–25,37], and higher incidence of EEV clinical cases was detected during outbreaks of AHS or other arboviruses in certain districts of South Africa [32]. Therefore, EEV surveillance may be important to infer on the circulation of other *Orbiviruses*, especially AHSV, against which many horses in southern Africa area are routinely vaccinated. The global changes in climate have led to changes in vectors' habitat and range and to the expansion in the habitat of *Culicoides* species, namely *C. imicola*, which is a major vector of EEV, AHSV, BTV, and other veterinary important *Orbiviruses*. These changes led to the emergence and spread of various *Orbiviruses* into more temperate regions and to an increase in global incidence and virus diversity [35,36,42]. Several studies aimed to evaluate the risk for the introduction of EEV into European countries pointed at two possible routes of introduction: importation of infected animals or importation of infected vectors, with the former being more probable [12,43,44]. In addition, the risk of the introduction of EEV through an infected host was higher than that of AHSV [12,43].

Herd immunity has also been suggested to serve as a protective factor against EEV infection. In Israel, lower annual incidence was recorded in farms with initial higher seroprevalence [6]. In South Africa, a possible pattern was suggested for predominant serotypes within an area in which high prevalence was recorded for a season or two, followed by a dramatic reduction in the incidence in the following year [22]. Immunity seems to be serotype specific, and maternal immunity (which is estimated to last until the age of 5–6 months) does not seem to prevent EEV infection of foals, both because of the variable composition of serotype-specific maternal antibodies and diminished maternal antibody levels prior to the high-risk season (the end of the rainy season) [14,22,30]. Transplacental transmission is probably not a major route of transmission, although abortions have been reported as a consequence of EEV infection, and the virus has been isolated from a placenta of a mare with a fatal case of EEV. Vertical transmission in the vertebrate host has been reported for BTV but not for AHSV [32].

EEV has been reported to infect various equid species, including horses, donkeys, and zebras in southern Africa, with similar prevalence rates and serotype distribution [8,23–25]. Donkeys are considered to be more resistant to clinical disease, are widely dispersed over various ecological zones, and are usually more susceptible to the presence of insect vectors. Therefore, donkeys are considered to be ideal sentinels for both EEV and AHSV [14,23]. A high-resolution study of zebras at Kruger National Park (KNP) demonstrated continued exposure to EEV throughout the year, attributed to the unbroken presence of the vector throughout the year in the subtropical climate. It has been suggested that zebras may play a role in the persistence and over-wintering of EEV in the area when *Culicoides* abundance is low due to the combination of high numbers of susceptible foals and sufficient numbers of *Culicoides* vectors during winter [24]. EEV had also been serologically detected in four elephants. This observation might be incidental or false (due to the non-specific reaction of elephant sera in serological test), and their role in the circulation of EEV in the area is yet undetermined [25].

4. Clinical Disease

The clinical significance of EEV infection is difficult to determine but is probably low. Generally, EEV is associated with mild or subclinical disease in horses with low mortality rates [3,4]. Characteristic clinical presentation of symptomatic horses consists of a short period (typically two to five days) of fluctuating fever and inappetence, sometimes accompanied by tachycardia and tachypnea [27]. Currently, there is no evidence of the zoonotic potential of EEV.

The name "equine encephalosis" is misleading, as the disease is not primarily neurologic. The name was given when the virus was first isolated from the organs of a mare that died during an outbreak at the farm in which three mares were affected, and two died. All three mares suffered from a peracute, febrile, nervous disease, and both fatal cases were diagnosed with edema and congestion of the brain, focal catarral enteritis, and mild fatty generation of the liver [2,14]. In the following years, EEV was isolated from horses exhibiting a variety of clinical signs including, fever, inappetence, central nervous system signs including severe ataxia, stiffness, changes in temperament and convulsions, respiratory signs including nasal discharge, enteritis, cardiac failure, liver damage and icterus, abortion (at 5–6 months), conjunctivitis, and swelling of the neck, lips or eyelids [4,9,11,14,28]. However, only limited numbers of clinical cases have been described, despite the high seroprevalence of EEV in South Africa suggesting that most cases are subclinical [30].

The clinical signs of EE are non-specific and could be easily confused with that of other viruses. Initially, EE symptoms were described as a mild manifestation of AHS, and EE outbreaks often coincide with outbreaks of AHS or other arboviruses. Therefore, to confirm the diagnosis of EEV as the cause of disease, the virus should be directly identified, and other potential pathogens should be ruled out [28,32]. In a recent analysis of 1523 samples from horses in South Africa presenting neurological, febrile, respiratory signs, or sudden death, 7.3% (111 horses) were infected with EEV (as diagnosed by real-time reverse-transcriptase PCR, rRT-PCR). Of these EEV-positive horses, 17 were co-infected with other arboviruses (AHSV, West Nile virus, or Middelburg virus). Clinical signs that were significantly associated with EEV-positive cases were fever, dyspnea, and icterus. In contrast, neurological signs (and specifically ataxia) and case fatality (including euthanasia) were inversely associated with EEV infection. Although 47.7% of EEV-positive horses had neurological abnormalities (some of which were co-infected with other viruses), only 9% had fatal outcomes [32]. In general, fatality rates following EEV infection are relatively low and estimated at 0% to 5% of clinical cases [4,9].

There is no sufficient data of possible associations between specific clinical signs and EEV serotypes. The "original" neurologic syndrome could only have been experimentally reproduced once, using the EEV-2 (Cascara) serotype [14]. Full genome sequences obtained from six clinical cases (three neurologic, one febrile, one dyspneic, and one abortion) classified five as EEV-1, while the horse with respiratory signs was infected with EEV-4 [32]. Since EEV-1 is the most prevalent serotype in South Africa [14,22], it is difficult to infer from these findings differences in pathogenicity between genotypes. In general, the negative association between neurological signs and case fatality and EEV in clinical cases [32] suggests that EEV is probably not a major cause of neurological disease or case fatality in endemic areas.

5. Treatment, Prevention, and Control

No specific treatment is available against EEV. Most symptomatic cases recover with no complications. Supportive treatment may be administered to decrease fever and inflammation or relief other clinical signs. No vaccine is currently available against EEV [4,27].

EEV is considered noncontagious, and prevention strategies mainly focus on reducing exposure to *Culicoides* vectors. Vector control is usually based on a combination of mechanical, chemical, biological, and genetic methods used to limit the vector's habitat and reduce vector-host contact. Such methods are most relevant to stabled horses and include stabling horses at dusk and dawn (when the vectors are more active), reducing light at night, screening windows, treating or removing animal waste, and using vector repellents on horses and the environment [4,45].

To prevent the introduction of EEV into new areas, transportation restrictions should be applied [45]. However, since EEV has limited veterinary and economic impact, such restrictions are not required to date. Modeling of the possibility of introduction of EEV from endemic countries into Europe demonstrated that control measures prior to exportation, including mostly quarantine and vector control, but also clinical inspection and serological screening, are efficient in reducing the probability of EEV introduction [12,43].

6. Diagnosis

Rapid and accurate diagnosis is important, especially during disease outbreaks, with sufficient specificity to distinguish between closely related pathogens in order to implement appropriate treatment and control. Different methods have been developed for the detection and classification of EEV, including virus isolation, serological assays, and molecular assays.

Historically, EEV was identified by virus isolation in baby hamster kidney (BHK) cells, suckling mice brain, embryonated chicken eggs, and Vero cells (African green monkey kidney cells) [2,4,11,19,29,32]. Most experimentally infected cell lines displayed post-infection cytopathic effects [11] and can be used for virus neutralization tests (VNT). However, virus isolation methods are labor intensive and are not very sensitive.

Several serological methods detecting anti-EEV antibodies have been developed, which are most useful for screening and epidemiological surveys. Some of these assays are group-specific enzyme-linked immunosorbent assays (ELISA) detecting antibodies against all EEV serotypes, but not of other *Orbiviruses.* These methods include competitive ELISA (cELISA) and indirect sandwich ELISA (S-ELISA), both having 100% sensitivity and specificity [46,47]. Other methods are serotype specific and used to determine EEV serotype, mainly VNT [14].

Molecular assays are usually very specific and sensitive and are increasingly being used for the identification and classification of EEV, as well as other *Orbiviruses*, at a serogroup and serotype level. Genomic probes have been developed for the detection of EEV. The NS1 (Seg-5) gene was the most sensitive for the detection of EEV at a serogroup level, while the VP2 (Seg-2) gene was serotype specific [18,48]. Real-time reverse-transcriptase polymerase chain reaction (rRT-PCR), using TaqMan probes have been developed to detect EEV at the serogroup level, using VP7 (seg-7) and VP6 (Seg-9) genes, and at the serotype level, using VP2 (seg-2) gene, with high sensitivity, specificity, and efficacy [19,38]. Full and partial genome sequencing (of one or more segments) of EEV have also been used for phylogenetic studies comparing viral species, isolates, or genotypes and are usually used for epidemiological investigations rather than routine clinical detection [11,20,27,32,49]. In some cases, next-generation sequencing was used to identify EEV from isolates or total RNA from clinical cases in new geographical areas, where EEV had not been suspected [9,11].

7. The Israeli Perspective

Between October 2008 and January 2009, a febrile horse disease was observed in hundreds of horses in more than 60 equine premises across Israel. Initial serological results indicated that the disease was equine viral arteritis (EVA), but this virus was not isolated, and PCR tests were all negative. Using a novel DNA array technique, with subsequent RT-PCR and sequence analysis in the Veterinary Laboratories Agency (VLA) in the United Kingdom (U.K.), the virus was identified as EEV [9]. This was the first time that this virus was isolated anywhere else north to South Africa. A year later, samples were collected from eight febrile horses in Israel, and cultures from three of these horses were positive for EEV [27]. Phylogenetic analysis of VP2 (Seg-2) showed 92% sequence identity to EEV-3, and the phylogenetic analysis of EEV NS3 (Seg-10) grouped these isolates with other EEV isolates but as a distinct group [27]. Based on these differences, it was speculated that this virus has evolved in the region for a sufficient time to accumulate these changes and was not recently introduced to Israel from South Africa [27]. Indeed, soon afterward, retrospective analysis of sera samples collected from horses in Israel for other reasons revealed anti-EEV antibodies in four of five sera samples that were collected in 2001 [28], and similar isolates have also been characterized in Gambia.

This sequence of events resembles the events leading to the first description of EEV in South Africa in 1967, in which the EEV outbreak was initially described as a mild case of AHS [2] (and later was connected to the first description from 1910 [1]), and its first isolation in India, following an outbreak of febrile disease that led to an investigation during which numerous other pathogens have been ruled out, and only next-generation sequencing of total RNA finally identified EEV as the cause of disease [11]. The course of these epidemiological investigations demonstrates the diagnostic challenge when encountering a newly introduced or emerging pathogen in a new area, especially when it does not have characteristic clinical signs.

During the 2008 outbreak in Israel in which more than 60 stables were affected, clinical signs included mainly: raised body temperature, pulse, and respiratory rates, unrest, and decreased appetite. Although morbidity was high (and reached 100% in some of the farms) and was reported in horses from different breeds, ages, and sexes, no fatalities were reported [9]. Since this outbreak, EEV has been sporadically diagnosed as a cause of febrile disease, with few local outbreaks. In these diagnosed or related cases, the clinical signs included fever and inappetence for 2–5 days, with complications of respiratory signs or colic in the minority of cases (unpublished data). This observation is contrary to other reports of severe neurological or life-threatening signs during EEV outbreaks [14,32]. It has been proposed that EEV serotypes may differ in their pathogenicity [14,32]; however, no evidence of such differences is currently available. In this respect, it is important to mention that during extensive outbreaks, it is probable that not all clinical cases are the result of the same pathogen, and unless definitively diagnosed, caution should be used when making such interpretations. Based on our experience, in an endemic area, where clinical cases are suspected every year, it seems that the clinical significance of EEV is low, and it is mainly important as a model for the spread of new pathogens to new niches where the vectors are present.

8. Conclusions

Although EEV is probably not of major clinical importance, its emergence in areas outside southern Africa may precede the spread of other, closely related, *Culicoides*-borne pathogens, such as AHSV and BTV. For the time being, the virus has been reported from southern and central Africa, Israel, and India but has not spread to moderate climate countries in Europe or the Americas. The lower virulence of this virus, in combination with uncharacteristic clinical signs, makes its diagnosis challenging, especially in areas when it is not known to be endemic. Future research is needed to better understand the epidemiology and pathogenesis of EEV, as well as the dynamics of the circulation of various *Culicoides*-borne arboviruses.

Author Contributions: Conceptualization, A.S. and S.T.-L.; investigation, S.T.-L. and A.S.; writing—original draft preparation, S.T.-L.; writing—review and editing, A.S. and S.T.-L.; visualization, S.T.-L.; All authors have read and agreed to the published version of the manuscript.

Funding: This research received no external funding.

Institutional Review Board Statement: Not applicable.

Informed Consent Statement: Not applicable.

Data Availability Statement: Not applicable.

Conflicts of Interest: The authors declare no conflict of interest.

References

1. Theiler, A. Notes on a fever in horses simulating horse-sickness. *Transvaal. Agric. J.* **1910**, *8*, 581–586.
2. Erasmus, B.; Adelaar, T.; Smit, J.; Lecatsas, G.; Toms, T. The isolation and characterization of equine encephalosis virus. *Bull. Off. Int. Epizoot.* **1970**, *74*, 781–789.
3. Attoui, H.; Jaafar, F.M. Zoonotic and emerging orbivirus infections. *Rev. Sci. Tech.* **2015**, *34*, 353–361. [CrossRef] [PubMed]
4. Dhama, K.; Pawaiya, R.; Karthik, K.; Chakrabort, S.; Tiwari, R.; Verma, A. Equine encephalosis virus (EEV): A Review. *Asian J. Anim. Vet. Adv.* **2014**, *9*, 123–133. [CrossRef]
5. Paweska, J.T.; Venter, G.J. Vector competence of Culicoides species and the seroprevalence of homologous neutralizing antibody in horses for six serotypes of equine encephalosis virus (EEV) in South Africa. *Med. Vet. Entomol.* **2004**, *18*, 398–407. [CrossRef]
6. Aharonson-Raz, K.; Steinman, A.; Kavkovsky, A.; Bumbarov, V.; Berlin, D.; Lichter-Peled, A.; Berke, O.; Klement, E. Analysis of the Association of Climate, Weather and Herd Immunity with the Spread of Equine Encephalosis Virus in Horses in Israel. *Transbound. Emerg. Dis.* **2015**, *64*, 593–602. [CrossRef]
7. MacLachlan, N.J.; Guthrie, A.J. Re-emergence of bluetongue, African horse sickness, and other Orbivirus diseases. *Vet. Res.* **2010**, *41*, 35. [CrossRef]
8. Gordon, S.J.; Bolwell, C.; Rogers, C.W.; Musuka, G.; Kelly, P.; Guthrie, A.; Mellor, P.S.; Hamblin, C. The sero-prevalence and sero-incidence of African horse sickness and equine encephalosis in selected horse and donkey populations in Zimbabwe. *Onderstepoort J. Vet. Res.* **2017**, *84*, 5. [CrossRef]
9. Mildenberg, Z.; Westcott, D.; Bellaiche, M.; Dastjerdi, A.; Steinbach, F.; Drew, T. Equine Encephalosis Virus in Israel. *Transbound. Emerg. Dis.* **2009**, *56*, 291. [CrossRef]
10. Oura, C.A.L.; Batten, C.A.; Ivens, P.A.S.; Balcha, M.; Alhassan, A.; Gizaw, D.; Elharrak, M.; Jallow, D.B.; Sahle, M.; Maan, N.; et al. Equine encephalosis virus: Evidence for circulation beyond southern Africa. *Epidemiol. Infect.* **2012**, *140*, 1982–1986. [CrossRef]
11. Yadav, P.D.; Albarino, C.G.; Nyayanit, D.A.; Guerrero, L.; Jenks, M.H.; Sarkale, P.; Nichol, S.T.; Mourya, D.T. Equine Encephalosis Virus in India. *Emerg. Infect. Dis.* **2018**, *24*, 898–901. [CrossRef] [PubMed]
12. Fischer, E.A.J.; López, E.P.M.; De Vos, C.J.; Faverjon, C. Quantitative analysis of the probability of introducing equine encephalosis virus (EEV) into The Netherlands. *Prev. Vet. Med.* **2016**, *131*, 48–59. [CrossRef] [PubMed]
13. King, A.M.; Adams, M.J.; Carstens, E.B.; Lefkowitz, E.J. Virus taxonomy. In *Ninth report of the International Committee on Taxonomy of Viruses*; Elsevier: London, UK, 2012.
14. Howell, P.G.; Groenewald, D.; Visage, C.W.; Bosman, A.-M.; Coetzer, J.A.W.; Guthrie, A.J. The classification of seven serotypes of equine encephalosis virus and the prevalence of homologous antibody in horses in South Africa. *Onderstepoort J. Vet. Res.* **2002**, *69*, 79–93. [PubMed]
15. Gould, A.R.; Hyatt, A.D. The orbivirus genus. Diversity, structure, replication and phylogenetic relationships. *Comp. Immunol. Microbiol. Infect. Dis.* **1994**, *17*, 163–188. [CrossRef]
16. Dennis, S.J.; Meyers, A.E.; Hitzeroth, I.I.; Rybicki, E.P. African Horse Sickness: A Review of Current Understanding and Vaccine Development. *Viruses* **2019**, *11*, 844. [CrossRef]
17. Huismans, H.; Van Staden, V.; Fick, W.C.; Van Niekerk, M.; Meiring, T.L. A comparison of different orbivirus proteins that could affect virulence and pathogenesis. *Vet. Ital.* **2010**, *40*, 417–425.
18. Viljoen, G.J.; Huismans, H. The Characterization of Equine Encephalosis Virus and the Development of Genomic Probes. *J. Gen. Virol.* **1989**, *70*, 2007–2015. [CrossRef]
19. Maan, S.; Belaganahalli, M.N.; Maan, N.S.; Potgieter, A.C.; Mertens, P.P.C. Quantitative RT-PCR assays for identification and typing of the Equine encephalosis virus. *Braz. J. Microbiol.* **2019**, *50*, 287–296. [CrossRef]
20. Van Niekerk, M.; Freeman, M.; Paweska, J.T.; Howell, P.G.; Guthrie, A.J.; Potgieter, A.C.; Van Staden, V.; Huismans, H. Variation in the NS3 gene and protein in South African isolates of bluetongue and equine encephalosis viruses. *J. Gen. Virol.* **2003**, *84*, 581–590. [CrossRef]
21. Lecatsas, G.; Erasmus, B.J.; Els, H.J. Electron microscopic studies on equine encephalosis virus. *Onderstepoort J. Vet. Res.* **1973**, *40*, 53–57.
22. Howell, P.G.; Nurton, J.P.; Nel, D.; Lourens, C.W.; Guthrie, A.J. Prevalence of serotype specific antibody to equine encephalosis virus in Thoroughbred yearlings South Africa (1999–2004). *Onderstepoort J. Vet. Res.* **2008**, *75*, 153–161. [CrossRef] [PubMed]

23. Lord, C.C.; Venter, G.J.; Mellor, P.S.; Paweska, J.T.; Woolhouse, M.E.J. Transmission patterns of African horse sickness and equine encephalosis viruses in South African donkeys. *Epidemiol. Infect.* **2002**, *128*, 265–275. [CrossRef] [PubMed]
24. Barnard, B.J.; Paweska, J.T. Prevalence of antibodies against some equine viruses in zebra (*Zebra burchelli*) in the Kruger National Park, 1991–1992. *Onderstepoort J. Vet. Res.* **1993**, *60*, 175–179. [PubMed]
25. Barnard, B.J. Antibodies against some viruses of domestic animals in southern African wild animals. *Onderstepoort J. Vet. Res.* **1997**, *64*, 95–110. [PubMed]
26. Williams, R.; Du Plessis, D.H.; Van Wyngaardt, W. Group-Reactive ELISAs for Detecting Antibodies to African Horsesickness and Equine Encephalosis Viruses in Horse, Donkey, and Zebra Sera. *J. Vet. Diagn. Investig.* **1993**, *5*, 3–7. [CrossRef] [PubMed]
27. Aharonson-Raz, K.; Steinman, A.; Bumbarov, V.; Maan, S.; Maan, N.S.; Nomikou, K.; Batten, C.; Potgieter, C.; Gottlieb, Y.; Mertens, P.; et al. Isolation and Phylogenetic Grouping of Equine Encephalosis Virus in Israel. *Emerg. Infect. Dis.* **2011**, *17*, 1883–1886. [CrossRef] [PubMed]
28. Wescott, D.G.; Mildenberg, Z.; Bellaiche, M.; McGowan, S.L.; Grierson, S.S.; Choudhury, B.; Steinbach, F. Evidence for the Circulation of Equine Encephalosis Virus in Israel since 2001. *PLoS ONE* **2013**, *8*, e70532. [CrossRef] [PubMed]
29. Tirosh-Levy, S.; Gelman, B.; Zivotofsky, D.; Quraan, L.; Khinich, E.; Nasereddin, A.; Abdeen, Z.; Steinman, A. Seroprevalence and risk factor analysis for exposure to equine encephalosis virus in Israel, Palestine and Jordan. *Vet. Med. Sci.* **2017**, *3*, 82–90. [CrossRef]
30. Grewar, J.D.; Thompson, P.N.; Lourens, C.W.; Guthrie, A.J. Equine encephalosis in Thoroughbred foals on a South African stud farm. *Onderstepoort J. Vet. Res.* **2015**, *82*, 4. [CrossRef]
31. Snyman, J.; Venter, G.J.; Venter, M. An Investigation of Culicoides (Diptera: Ceratopogonidae) as Potential Vectors of Med-ically and Veterinary Important Arboviruses in South Africa. *Viruses* **2021**, *13*, 1978. [CrossRef]
32. Snyman, J.; Koekemoer, O.; van Schalkwyk, A.; van Vuren, P.J.; Snyman, L.; Williams, J.; Venter, M. Epidemiology and Genomic Analysis of Equine Encephalosis Virus Detected in Horses with Clinical Signs in South Africa, 2010–2017. *Viruses* **2021**, *13*, 398. [CrossRef] [PubMed]
33. Venter, G.J.; Groenewald, D.; Venter, E.; Hermanides, K.G.; Howell, P.G. A comparison of the vector competence of the biting midges, Culicoides (Avaritia) bolitinos and C. (A.) imicola, for the Bryanston serotype of equine encephalosis virus. *Med. Vet. Èntomol.* **2002**, *16*, 372–377. [CrossRef] [PubMed]
34. Venter, G.J.; Groenewald, D.M.; Paweska, J.T.; Venter, E.; Howell, P.G. Vector competence of selected South African Culicoides species for the Bryanston serotype of equine encephalosis virus. *Med. Vet. Èntomol.* **1999**, *13*, 393–400. [CrossRef] [PubMed]
35. Purse, B.V.; Carpenter, S.; Venter, G.J.; Bellis, G.; Mullens, B.A. Bionomics of Temperate and Tropical Culicoides Midges: Knowledge Gaps and Consequences for Transmission of Culicoides-Borne Viruses. *Annu. Rev. Èntomol.* **2015**, *60*, 373–392. [CrossRef] [PubMed]
36. Purse, B.V.; Mellor, P.S.; Rogers, D.J.; Samuel, A.R.; Mertens, P.; Baylis, M. Climate change and the recent emergence of bluetongue in Europe. *Nat. Rev. Microbiol.* **2005**, *3*, 171–181. [CrossRef] [PubMed]
37. Venter, G.J.; Koekemoer, J.J.O.; Paweska, J.T. Investigations on outbreaks of African horse sickness in the surveillance zone in South Africa. *Rev. Sci. Tech.* **2006**, *25*, 1097–1109. [CrossRef] [PubMed]
38. Rathogwa, N.; Quan, M.; Smit, J.; Lourens, C.; Guthrie, A.; Van Vuuren, M. Development of a real time polymerase chain reaction assay for equine encephalosis virus. *J. Virol. Methods* **2014**, *195*, 205–210. [CrossRef]
39. Mellor, P.S.; Boorman, J.; Baylis, M. Culicoides Biting Midges: Their Role as Arbovirus Vectors. *Annu. Rev. Èntomol.* **2000**, *45*, 307–340. [CrossRef]
40. Baylis, M.; Mellor, P.S.; Meiswinkel, R. Horse sickness and ENSO in South Africa. *Nature* **1999**, *397*, 574. [CrossRef]
41. Calzolari, M.; Albieri, A. Could drought conditions trigger Schmallenberg virus and other arboviruses circulation? *Int. J. Health Geogr.* **2013**, *12*, 7. [CrossRef]
42. Leta, S.; Fetene, E.; Mulatu, T.; Amenu, K.; Jaleta, M.B.; Beyene, T.J.; Negussie, H.; Revie, C.W. Modeling the global distribution of Culicoides imicola: An Ensemble approach. *Sci. Rep.* **2019**, *9*, 14187. [CrossRef]
43. Faverjon, C.; Leblond, A.; Lecollinet, S.; Bødker, R.; De Koeijer, A.A.; Fischer, E. Comparative Risk Analysis of TwoCulicoides-Borne Diseases in Horses: Equine Encephalosis More Likely to Enter France than African Horse Sickness. *Transbound. Emerg. Dis.* **2016**, *64*, 1825–1836. [CrossRef] [PubMed]
44. Zimmerli, U.; Herholz, C.; Schwermer, H.; Hofmann, M.; Griot, C. African horse sickness and equine encephalosis: Must Switzerland get prepared? *Schweiz. Arch. Tierh.* **2010**, *152*, 165–175. [CrossRef] [PubMed]
45. Harrup, L.; Miranda, M.; Carpenter, S. Advances in control techniques for Culicoides and future prospects. *Vet. Ital.* **2016**, *52*, 247–264. [CrossRef] [PubMed]
46. Crafford, J.; Guthrie, A.; Van Vuuren, M.; Mertens, P.; Burroughs, J.; Howell, P.; Batten, C.; Hamblin, C. A competitive ELISA for the detection of group-specific antibody to equine encephalosis virus. *J. Virol. Methods* **2011**, *174*, 60–64. [CrossRef]
47. Crafford, J.; Guthrie, A.; Van Vuuren, M.; Mertens, P.; Burroughs, J.; Howell, P.; Hamblin, C. A group-specific, indirect sandwich ELISA for the detection of equine encephalosis virus antigen. *J. Virol. Methods* **2003**, *112*, 129–135. [CrossRef]
48. Venter, E.; Viljoen, G.; Nel, L.; Huismans, H.; Van Dijk, A. A comparison of different genomic probes in the detection of virus-specified RNA in Orbivirus-infected cells. *J. Virol. Methods* **1991**, *32*, 171–180. [CrossRef]
49. Quan, M.; van Vuuren, M.; Howell, P.G.; Groenewald, D.; Guthrie, A.J. Molecular epidemiology of the African horse sickness virus S10 gene. *J. Gen. Virol.* **2008**, *89*, 1159–1168. [CrossRef]

MDPI
St. Alban-Anlage 66
4052 Basel
Switzerland
Tel. +41 61 683 77 34
Fax +41 61 302 89 18
www.mdpi.com

Animals Editorial Office
E-mail: animals@mdpi.com
www.mdpi.com/journal/animals

www.ingramcontent.com/pod-product-compliance
Lightning Source LLC
LaVergne TN
LVHW070618170726
843515LV00004B/121
* 9 7 8 3 0 3 6 5 5 0 8 6 2 *